FIGURES OF LIGHT

By Stanley Kauffmann

Novels

The Hidden Hero
The Tightrope
A Change of Climate
Man of the World

Children's Play

Bobino

Criticism

A World on Film
Figures of Light

Stanley Kauffmann

FIGURES OF LIGHT
film criticism and comment

HARPER & ROW, PUBLISHERS
NEW YORK, EVANSTON, SAN FRANCISCO, LONDON

1817

Grateful acknowledgment is given to the following:

Channel 13/WNDT for *"Falstaff"* from the program "The Art of Film" produced on April 5, 1967.
Salamagundi for "A Year with *Blow-Up:* Some Notes" appearing in the Spring-Summer 1968 issue.
The New York Times for *"Who's Afraid of Virginia Woolf?"* Copyright © 1966 by The New York Times Company. Reprinted by permission.
Liveright Publishing Corp. for the lines on page 226 from "Chaplinesque" from *The Complete Poems and Selected Letters and Prose of Hart Crane.* Copyright 1933, © 1958, 1966 by Liveright Publishing Corp., New York. Reprinted by permission.

FIRST EDITION

LIBRARY OF CONGRESS CATALOG CARD NUMBER: 70-138742

to Laura

contents

preface

At the moment I'm still alive. I mention the fact a bit defiantly because an English reviewer of my first collection of film criticism expressed gratitude for James Agee's and Robert Warshow's posthumous collections but indicated that it was remiss of me to publish a collection while I was still alive, instead of writing a "real" book. And here I am offering a second collection without having had the decency to expire.

Real or otherwise, this book was written, week by week: not as serial installments of one work, certainly, yet it is a (selective) record of a continuing relation with an art.

The first volume of that record, *A World on Film,* consisted mainly of my criticism in *The New Republic* from early 1958 to November 1965. In 1966 I spent eight months as drama critic of *The New*

York Times, where I also wrote a few film reviews, such as the first piece in this book. Late in 1966 I returned to *The New Republic* as weekly book critic, during which time I wrote film criticism for the *New American Review,* and some of that, too, is included here. In late 1967 I resumed as film critic of *The New Republic;* most of the following material comes from that source. Other sources are specified, except for the last selection, which is drawn from an article written for *Quality: Its Image in the Arts,* edited by Louis Kronenberger and published by Atheneum. As in my first book, I have occasionally added postscripts.

But the arrangement of this book differs from the earlier one, and that difference is pertinent to the idea of a continuing record. The reviews in *A World on Film* were arranged by country of origin and by topic because, for me, the dominant impression of that period was of simultaneity of response around the film-making world. In the years since then, the dominant impression has become one of the speed of change—in film-making styles and materials. Any arrangement of this book other than chronological seemed to me to fight that dominant idea; so these pieces are presented in the order in which they were written, which is almost always the order in which they were published.

One point is the same as in the first book, and I'd like simply to quote from that preface. "I cannot imagine a more stimulating life for a critic of new works than to be able to address regularly a group of the best readers he can think of and to be given a free hand in doing it." Gratitude, again, to the readers of *The New Republic* and to its editor, Gilbert Harrison.

S.K.

August, 1970

FIGURES OF LIGHT

Who's Afraid of Virginia Woolf?

(*The New York Times, June 24, 1966*)

EDWARD Albee's *Who's Afraid of Virginia Woolf?*, the best American play of the last decade and a violently candid one, has been brought to the screen without pussyfooting. This in itself makes it a notable event in our film history. About the film as such, there is more to be said.

First things first. The most pressing question—since we already know a great deal about the play and the two stars—is the direction. Mike Nichols, after a brilliant and too brief career as a satirist, proved to be a brilliant theatrical director of comedy. This is his debut as a film director, and it is a successful Houdini feat. Houdini, you remember, was the magician who was chained hand and foot, bound in a sack, dumped in a river, and then appeared some minutes later on the surface. You do not expect Olympic swimming form in a Houdini; the triumph is just to come out alive.

Which Nichols has done. He was given two world-shaking stars, the play of the decade and the auspices of a large, looming studio. What more inhibiting conditions could be imagined for a first film, if the director is a man of talent? But Nichols has at least survived. The form is not Olympic, but he lives.

Any transference of a good play to film is a battle. (That is why the best film directors rarely deal with good plays.) The better the play, the harder it struggles against leaving its natural habitat, and Albee's extraordinary comedy-drama has put up a stiff fight. Ernest Lehman, the screen adapter, has broken the play out of its one living-room setting into various rooms in the house and onto the lawn, which the play accepts well enough. He has also placed one scene in a roadhouse, which is a patently forced move for visual variety. These changes and some minor cuts, including a little inconsequential blue-penciling, are about the sum of his efforts. The real job of "filmizing" was left to the director.

With no possible chance to cut loose cinematically (as, for example, Richard Lester did in his film of the stage comedy *The Knack*), Nichols has made the most of two elements that were left to him— intimacy and acting. He has gone to school to several film masters (Kurosawa among them, I would guess) in the skills of keeping the camera close, indecently prying; giving us a sense of his characters'

1

very breath, bad breath, held breath; tracking a face—in the rhythm of the scene—as the actor moves, to take us to other faces; punctuating with sudden withdrawals to give us a brief, almost dispassionate respite; then plunging us in close again to one or two faces, for lots of pores and bile.

There is not much that is original in Nichols' camerawork, no sense of the personality that we got in his stage direction. In fact, the direction is weakest when he gets a bit arty: electric signs flashing behind heads or tilted shots from below to show passion and abandon (both of them hallmarks of the college cinema virtuoso). But he has minimized the "stage" feeling, and he has given the film an insistent presence, good phrasing, and a nervous drive. It sags toward the end, but this is because the third act of the play sags.

As for the acting, Nichols had Richard Burton as George. To refresh us all, George is a fortyish history professor, married to Martha, the daughter of the president of a New England college. They return home from a party at 1:30 A.M., slightly sozzled, drenched in their twenty-year-old marital love-hate ambivalence. A young faculty couple come over for drinks, and the party winds viciously on until dawn. In the course of it, Martha sleeps with the young man as an act of vengeance on George. The play ends with George's retribution—the destruction of their myth about a son they never had.

Burton was part of the star package with which this film began, but—a big but—Burton is also an actor. He has become a kind of specialist in sensitive self-disgust, as witness the latter scenes of *Cleopatra* and all of *The Spy Who Came In From the Cold,* and he does it well. He is not in his person the George we might imagine, but he is utterly convincing as a man with a great lake of nausea in him, on which he sails with regret and compulsive amusement.

On past evidence, Nichols had relatively little work to do with Burton. On past evidence, he had a good deal to do with Elizabeth Taylor, playing Martha. She has shown previously, in some roles, that she could respond to the right director and could at least flagellate herself into an emotional state (as in *Suddenly, Last Summer*). Here, with a director who knows how to get an actor's confidence and knows what to do with it after he gets it, she does the best work of her career, sustained and urgent.

Of course, she has an initial advantage. Her acceptance of gray

hair and her use of profanity make her seem to be acting even (figuratively) before she begins. ("Gee, she let them show her looking old!
Wow, she just said 'Son of a bitch'! A star!") It is not the first time
an American star has gotten mileage out of that sort of daring. Miss
Taylor does not have qualities that, for instance, Uta Hagen had in
the Broadway version, no suggestion of endlessly coiled involutions.
Her venom is nearer the surface. But, under Nichols' hand, she
gets vocal variety, never relapses out of the role, and she charges
it with the utmost of her powers—which is an achievement for any
actress, great or little.

As the younger man, George Segal gives his usual good terrier performance, lithe and snapping, with nice bafflement at the complexities
of what he thought was simply a bad marriage. As his bland wife,
Sandy Dennis is credibly bland.

Albee's play looks both better and a little worse under the camera's
magnification. A chief virtue for me is that it is not an onion skin
play—it does not merely strip off layers, beginning at the surface
with trifles and digging deeper as it proceeds. Of course, we learn
more about the characters as we go, and almost all of it is fascinating;
but, like its giant forebear, Strindberg's *Dance of Death,* the play
begins in hell, and all the revelations and reactions take place within
that landscape.

What does not wear well in the generally superb dialogue is the
heavy lacing of vaudeville cross-talk, particularly facile non sequi-
turs. (Also, in Lehman's version, so much shouting and slamming
takes place on the front lawn at four in the morning that we keep
wondering why a neighbor doesn't wake up and complain.)

More serious is the heightened impression that the myth of the
son is irrelevant to the play. It seems a device that the author tacked
on to conclude matters as the slash and counterslash grew tired;
a device that he then went back and planted earlier. Else why would
Martha have told the other woman the secret of the son so glibly—not
when she was angry or drunk—if she knew she was breaching an
old and sacred compact with her husband? It obtrudes as an arbitrary
action to justify the ending.

The really relevant unseen character is not the son; it is Martha's
father, the president of the college. It is he whom she idolizes and
measures her husband against, it is his presence George has to con-

tend with in and out of bed. It is Daddy's power, symbolic in Martha, that keeps the visiting couple from leaving, despite circumstances that would soon have driven them out of any other house.

Awareness—of this truth about Daddy, of multiple other truths about themselves and their world—is the theme of this play: not the necessity of narcotic illusion about the son, but naked, peeled awareness. Under the vituperation and violence, under Martha's aggressive and self-punishing infidelities, this is the drama of a marriage flooded with more consciousness than the human psyche is at present able to bear.

Their world is too much with them, their selves are much too clear. It is the price to be paid for living in a cosmos of increasing clarity—which includes a clearer view of inevitable futilities. And, fundamentally, it is this desperation—articulated in a childless, broken-hearted, demonically loving marriage—that Albee has crystallized in his flawed but fine play.

And in its forthright dealing with the play, this becomes one of the most scathingly honest American films ever made. Its advertisements say "No one under 18 will be admitted unless accompanied by his parent." This may safeguard the children; the parents must take their chances.

Falstaff

(WNDT-TV, New York, April 5, 1967)

ONE of the most brilliant directors in film history has produced another disappointment in an increasingly disappointing career. Orson Welles's Falstaff is a talented disaster. Certainly it is not without the dazzling directorial touches that are in everything he does (particularly the battle scenes), but most of those touches (excepting the battle scenes) are quite of the wrong, tristful, sere-and-yellow-leaf variety.

Welles's failure here as director derives, I believe, from his laziness as actor. In his films of Othello and Macbeth, apart from their other inadequacies, he refused the major challenges in the title roles. He fudged the big moments, found clever ways around them, gave us glimpses instead of confrontations, skimpy modern understatement instead of full-throated classical technique. In *Falstaff,* with a script puréed from both parts of *Henry IV* and from *Henry V,* Welles is simply too lazy—or perhaps by now too frightened—to take on the role as Shakespeare wrote it, not a jolly saloon-mural fat man but a giant of slyness, wit, guile, and vitality. W. H. Auden writes that Falstaff says, in effect, "I am I. Whatever I do, however outrageous, is of infinite importance because I do it."

But this ebullience, this radiant self-confident energy, would be too much work for Welles, who has let the muscles of his talent get flabby, so again he devises a way *around* the part. He erects a theory of Falstaff as sinning saint or somnolent pseudo-Hamlet, which makes the character very much less trouble for him to play.

We all know that he made this film, as he made several of his previous films, under great financial difficulties: in bits and pieces as he could scrape up the capital. Laurels to his persistence, then, but these stories of the business trials only make the result more rueful. All that business enterprise only to perpetuate an artistic laziness.

And willfulness. What economic necessity forced him to use Jeanne Moreau as Doll Tearsheet? Here, putting the Left Bank on the South Bank, he has miscast Miss Moreau more flagrantly than he did in *The Trial.* The story of Welles's career begins to look like one of the great wastes of talent in our time. His films now give off a whiff of surly wrongheadedness, perhaps as his anger at self-wastage and at wastage by others begins to harden into an intricate but poorly reasoned *position.*

A Year with *Blow-Up*—Some Notes

(*Salmagundi, Spring 1968*)

I first saw *Blow-Up* in early December 1966, and in the months since, I have broadcast, written, and lectured considerably about it. Here are some notes on the experience.

Quite apart from its intrinsic qualities, *Blow-Up* is an extraordinary social phenomenon. It is the first film from abroad by a major foreign director to have immediate national distribution. It was seen here more widely and more quickly than, for instance, *La Dolce Vita* for at least two reasons: it was made in English and it was distributed by a major American company. There are other reasons, less provable but probably equally pertinent: its mod atmosphere, its aura of sexuality, and, most important, its perfect timing. The end of a decade that had seen the rise of a film generation around the country was capped with a work by a recognized master that was speedily available around the country.

So this was the first time in my experience that a new film had been seen by virtually everyone wherever I talked about it. Usually the complaint had been (by letter) after a published or broadcast review, "Yes, but where can we see the picture?" Or, after a talk at some college not near New York, "But it will take years to get here, if ever," or "We'll have to wait until we can rent a 16-millimeter print." With *Blow-Up*, people in Michigan and South Carolina and Vermont knew—within weeks of the New York premiere—the film that was being discussed. This exception to the usual slow-leak distribution of foreign films had some interesting results.

A happy result was that people had seen this picture at the local Bijou. Before, many of them had seen Antonioni only in film courses and in film clubs. This one they had seen between runs of *How to Steal a Million* and *Hombre*. To those in big cities this may seem commonplace, but in smaller communities it was a rare event and had some good effects. To some degree it alleviated culture-vulturism and snobbism; *everyone* in Zilchville had seen *Blow-Up*, not just the elite; so happily there was no cachet simply in having seen it. Further, the fact of seeing it at the Bijou underscored those elements in the film medium of popular mythos that are valuable and valid—all those undefined and undefinable powers of warm communal embrace in the dark.

But there were some less happy results of the phenomenon. There was a good deal of back-formation value judgment. Because *Blow-Up* was a financial success, it could not really be good, I heard, or at least it proved that Antonioni had sold out. We heard the same thing about Bellow and *Herzog* when that book became a best seller. The parallel holds further in that *Blow-Up* and *Herzog* seem to me flawed but utterly uncompromised works by fine artists. I confess I got a bit weary of pointing out that to condemn a work because it is popular is exactly as discriminating as praising it because it is a hit.

Another discouraging consequence. Much of the discussion reflected modes of thought inculcated by the American academic mind, particularly in English departments. Almost everywhere there were people who wanted to discuss at length whether the murder in *Blow-Up* really happened or was an illusion. Now Antonioni, here as before, was interested in ambiguities; but ambiguities in art, like those in life, arise only from unambiguous facts—which is what makes them interesting. Anna in *L'Avventura* really disappeared; the ambiguities in morality arise from that fact. The lover in *Blow-Up* was really killed; the ambiguities in the hero's view of experience would not arise without that fact. (A quick "proof" that the murder was real: If it were not, why would the girl have wanted the pictures back? Why would the photographer's studio have been rifled? Why would we have been shown the pistol in the bushes? Some have even suggested that the corpse we see is a dummy, or a live man pretending. But Hemmings touches it.) This insistence on anteater nosings in the film seemed much less a reflection on *Blow-Up* than on an educational system—a system that mistakes factitious chatter for analysis.

On the other hand, an art teacher in a Nashville college told me, while driving me to the airport, that *Blow-Up* had given him a fulcrum with which to jimmy his previously apathetic students into *seeing:* seeing how the world is composed, how it is taken apart and recomposed by artists. In his excitement he almost drove off the road twice.*

The script is by Antonioni and his long-time collaborator Tonino Guerra, rendered in English by the young British playwright Edward Bond. It was suggested by a short story of the same name by the Argentinian author, resident in France, Julio Cortázar. Comparison of script and story is illuminating.

* In the *Saturday Review* of December 27, 1969, Larry Cohen, recently graduated from the University of Wisconsin, speaks of the "new audience" and its interest in relevance: "In this regard, *Blow-Up* functioned as the pivotal film; it radicalized the way in which many college students responded to film. More than three years after it first appeared, it remains a significant milestone in this country's awareness of motion pictures."

In the story the hero is an Argentine translator living in Paris—only an amateur photographer. One day while out walking, he photographs what he thinks is a pickup—of a youth by an older woman. There is an older man sitting in a parked car nearby. After the picture is taken, the youth flees, the woman protests, and the man in the car gets out and protests, too. Later, studying the picture, the hero sees (or imagines he sees) that the woman was really procuring the youth for the man in the car and that the fuss over the photograph gave the youth a chance to escape. It is a story of the discovery of, in Cortázar's view, latent horror, the invisible immanence of evil. (It is incidentally amusing that the photographer feels no such horror when he thinks that the woman is seducing the boy for herself.)

Antonioni retains little other than the device of subsequently discovering in a photograph what was really happening at that moment. He makes the hero a professional photographer, thus greatly intensifying the meaning of the camera in his life. By changing from the presumption of homosexuality to the fact of murder, Antonioni not only makes the discovered event more potent dramatically, he shifts it morally from the questionable to the unquestionable. At any time in history homosexuality has varied, depending on geography, in shades of good and evil. Murder, though more blinkable at some times and in some places than others, will be an evil fact so long as life has value.

Most important, Antonioni shifts the moral action from *fait accompli* to the present. His hero does not discover that he has been an agent of good, in a finished action. His dilemma is now.

With his recent films Antonioni has suffered, I think, from two professional failings of critics. The first has been well described in a penetrating review of *Blow-Up* by Robert Garis (*Commentary,* April 1967). Garis notes that Henry James's public grew tired of him while he was inconsiderate enough to be working out his career and sticking to his guns; that Beckett, after the establishment of *Waiting for Godot* as a masterpiece,

has been writing other beautiful and authentic plays quite similar to *Godot,* innocently unaware of that urgent necessity to move on, to find new themes and styles, that is so obvious to some of his critics . . . If it is regrettable to see the public wearing out new fashions in art as fast as automobiles, it is detestable to see criticism going along with this, if not actually leading the charge. The Antonioni case is like Beckett's but intensified. There has been the same puzzled annoyance with an artist who keeps on thinking and feeling about themes that everyone can see are worn out—themes like "lack of communication" or "commitment." There has been the same eagerness to master a difficult style and then the same relapse into boredom when that style turns out to be something the artist really takes seriously because that's the way he really sees things.

Another point grows out of this. This impatience with artists who are less interested in novelty than in deeper exploration leads to critical blindness about subtle gradations *within* an artist's "territory." We saw a gross example of this blindness last year in the theater when Harold Pinter's *The Homecoming* was shoved into the "formless fear" bag along with his earlier plays. The fact that Pinter had shifted focus, that he was now using his minute, vernacular, almost Chinese ritualism to scratch the human cortex for comic purposes, not for frisson—this was lost on most reviewers, who were just feeling comfy at finally having "placed" Pinter.

So with Antonioni. He's the one who deals with alienation and despair, isn't he? So the glib or the prejudiced have the pigeonhole all ready. Obviously the temper of Antonioni, like that of any genuine artist, is bound to mark all his work; but even in his last film, *Red Desert,* as it seemed to me, he had pushed into new areas of his "territory," was investigating the viability of hope, and had—without question—altered the rhythms of his editing to underscore a change of inquiry (not of belief). The editing is altered even further in *Blow-Up.* For instance, the justly celebrated sequence in which the hero suspects and then finds the murder in the photograph is quite unlike anything Antonioni has done before, in its accelerations and retards within a cumulative pattern. And the theme, too, seems to me an extension, a fresh inquiry, within Antonioni's field of interest. Here his basic interest seems to be in the swamping of consciousness by the conduits of technology. The hero takes some photographs of lovers, and thinks he has recorded a certain experience of which he is conscious; but, as he learns subsequently, his technology has borne in on him an experience of which he was not immediately aware, which he cannot understand or handle. He is permanently connected with a finished yet permanently unfinished experience. It seems to me a good epitome of same-size man vis-à-vis the expanding universe.

There are concomitant themes. One of them is success—but success *today,* which is available to youth as it has never been. The hero has money, and the balls that money provides in a money society, about twenty years earlier than would have been the general case twenty years ago; and he is no rare exception. Yet his troubles in this film do not arise from his money work but out of his "own" work, the serious work he does presumably out of the stings of conscience. (What else would drive a fashion-world hero to spend the night in a flophouse?) He can handle the cash cosmos; it is when he ventures into himself, leaves commissioned work and does something of his own, that he gets into trouble.

Along with this grows the theme of youth itself. This world is not only filled with but dominated by youth, in tastes and tone. (There are only two non-youths of prominence in the film. One, a nasty old clerk in an antique shop, resents the youth of the hero on sight. The other, a middle-aged lover, gets murdered.) The solidarity of youth is demonstrated in the hero's compunction to "prove" his strange experience to *his friends.* When one of his friends, the artist's wife, suggests that he notify the police of the murder (and her suggestion is in itself rather diffident), he simply doesn't answer to the point, as if law and criminality were outside the matter to him. Ex-soldiers say they can talk about combat only with other ex-soldiers. The communion of generations is somewhat the same. The hero doesn't want Their police; he wants certification by his friends.

Color is exquisite in Antonioni's films, and it is more than décor or even commentary; it is often chemically involved in the scene. In the shack in *Red Desert,* the walls of the bunk in which the picnickers lounge are bright red and give a highly erotic pulse to a scene in which sex is only talked about. In *Blow-Up* the hero and two teen-age girls have a romp on a large sheet of pale purple-lavender paper that cools a steamy little orgy into a kind of idyll.

This is the first feature that Antonioni has made outside Italy, and it shows a remarkable ability to cast acutely in a country where he does not know the corps of working actors intimately. (He discovered David Hemmings in the Hampstead Theatre Club "off-Broadway.") It also shows a remarkable ability to absorb and redeploy the essences of a foreign city without getting either prettily or grimly picturesque. But there are three elements in this film that betray some unease—an unease attributable perhaps to the fact that he was "translating" as he went, not only in language but also in experience.

The first is the plot strand of the neighboring artist and his wife. (Several people asked during the year why I call her his wife and not his girl friend. Answer: She wears a wedding ring. If it is unduly naïve to assume from this that the artist is her husband, it seems unduly sophisticated to assume that he is not.) This element has the effect of patchwork, as if it had not been used quite as intended or as if it were unfulfilled in its intent. The relationship between them and the hero is simply not grasped. The poorest scene in the entire film is the one in which the wife (Sarah Miles) visits the hero's studio after he has seen her making love with her husband. It is wispy and scrappy. The discomfort is the director's.

The second is the scene with the folk-rock group and the stampede for the discarded guitar. The scene has the mark of tourism on it, a phenomenon observed by an outsider and included for completeness' sake. Obviously Antonioni would not be a member of such a group in Italy any more than in England, but he would have known a thousand subtle things about Italian youths and their backgrounds that might have made them seem particularized, less a bunch of representatives *en bloc*.

The third questionable element is the use of the clown-faced masquers at the beginning and the end—which really means at the end, because they would not have been used at the start except to prepare for the end. Firstly, the texture of these scenes is jarring. Their symbolism—overt and conscious—conflicts with the digested symbolism of the rest of the film. It has a mark of strain and unfamiliarity about it, again like a phenomenon observed (partying Chelsea students, perhaps) and uncomfortably adapted. It is Cocteau strayed into Camus.

Much has been made of the clowns' thematic relevance, in that they provide a harbor of illusion for the hero after a fruitless voyage into reality. But precisely this thematic ground provides an even stronger objection to them, I think, than the textural one mentioned above. Thematically I think that the film is stronger without them, that it makes its points more forcibly. Suppose the picture began with Hemmings coming out of the flophouse with the derelicts, conversing with them, then leaving them and getting into his Rolls. At once it seems more like Antonioni. And suppose it ended (where in fact I thought it was going to end) with the long shot of Hemmings walking away after he has discovered that the corpse has been removed. Everything that the subsequent scene supplies would already be there by implication—*everything*—and we would be spared the cloudy symbols of high romance. Again it would be more like Antonioni.

All three of these lapses can possibly be traced to his working in a country where every last flicker of association and hint is not familiar and subconsciously secure.

On the other hand I think that two much-repeated criticisms of *Blow-Up* are invalid. Some have said that Antonioni seemingly ridicules the superficial world of fashion but is really reveling in it, exploiting it. It is hard to see how he could have made a film on this subject without photographing it. One may as well say that, in exploring the world of sexual powers and confusions in *A Married Woman*, Godard merely exploited nudity. By showing us the quasi-tarts of fashion (sex appeal, instead of sex, by the hour) in all their gum-chewing vapidity and by showing us how easily the Superbeautiful can be confected by someone who understands beauty, Antonioni does more than mock a conspicuously consuming society: he creates a laughable reality against which to pose a genuinely troubling ambiguity.

Another widespread objection has been to the role that Vanessa Redgrave plays. The character has been called unclear. But this seems to me true only in conventional nineteenth-century terms of character development. In a television interview with me, Antonioni said that Miss Redgrave read the script and wanted to play the part because— he lifted his hand in a gesture of placement on the screen—"*Sta lì.*"

"She stands there"; she has no explanations, no antecedents, no further consequences in the hero's life. I take this to mean that she is an analogue of the murder itself, an event rather than a person, unforgettable yet never knowable, and therefore perfectly consonant with the film.

Purely professionally from an actress's view, the role was a challenge because she has only two scenes, and those relatively brief appearances have to be charged with presence at once satisfying and tantalizing. Miss Redgrave met this challenge with ease, I think, not only because she has beauty and personality and distinctive talent but because she played the role against an unheard counterpoint: a secret and complete knowledge of who this young woman is. There is almost a hint that she is protecting Hemmings, that if he knew all that she knows, his life would disintegrate.

Thus my year of recurrent involvement with *Blow-Up* and some observations about it. Now Antonioni has announced that he will make a film in America. I have two feelings about this. I am glad: because I would like to see how he sees America, just as I was glad to see his view of London. I also have reservations: because he functions more completely where he is rooted. Fine directors like Renoir and Duvivier and Seastrom and Eisenstein have all made films away from home, all of which contain good things but none of which is that man's best work. Some directors, like Lean and Huston, have functioned at their best in foreign places, but they are not quintessentially societal directors. Antonioni (like Bergman in this respect) has been best in a society that is second nature to him, that has long fed and shaped him, that he has not had to "study."

It is something of a miracle that he made *Blow-Up* as well as he did. Its imperfections arise, I think, from having to concentrate on the miracle. Still, to say that I like *Blow-Up* the least of his films since *L'Avventura* is a purely relative statement. I would be content to see one film a year as good as *Blow-Up*—from Antonioni or anyone else—for the rest of my life.

Persona

(*New American Review #2, January 1968*)

SHORTLY after I saw Ingmar Bergman's *Persona* for the first time, I discovered the writings of R. D. Laing. Laing is a Scottish psychiatrist, blazingly humane, who is trying to understand (among other things) how madness becomes the sanity of the mad. A passage from his book *The Divided Self* might serve as epigraph for *Persona:*

The unrealness of perceptions and the falsity and meaninglessness of all activity are the necessary consequences of perception and activity being in the command of a false self—a system partially dissociated from the "true" self. . . .

Bergman's film begins with an actress, young and successful, who has suffered these consequences. All activity has become false and meaningless to her. In the middle of a performance of *Electra* she stops cold; after a moment she giggles and continues; but next day she is mute, almost catatonic. Her physician understands that the actress is obsessed by this disjunction between true and false selves. The actress is shackled, says the doctor, by a sense of her own false-ness, by the growing difference between what the world thinks of her and what she knows of herself. (Her profession is well chosen by Bergman to underscore her condition: not because she pretends nightly to be someone else but because the truth she creates in that pretense is greater than the truth of her own being.) What can one do? the doctor says calmly to the mute patient. Suicide? No, that would be vulgar. Silence. In ceasing to speak, the actress has, by her own values at least, stopped lying.

The doctor assigns to her a nurse of about her age, then sends patient and nurse to the doctor's own country house by the sea. There, in the dissolved-pearl light of the Swedish coast that we saw in *Through a Glass Darkly,* most of *Persona* takes place.

Before the titles of the film, we see a series of disconnected shots: a film projector's arc light hissing alive into glare, the "leader" of a reel of film, a snatch of silent-film slapstick, a sheep's eye being gouged, a nail being driven through a human hand. In a morgue full of corpses an adolescent boy sits up, reads, then stretches his hand to-ward a pane of glass behind which is an immense face that soon melds with another immense face. Later we learn that these are the faces of the actress (his mother?) and of the nurse. This dis-jointed beginning, made of splinters of horror and showmanship, is like a quick jagged tour of the actress's mind—images that terrify and also, in an Olympian way, amuse her.

After the titles, the film slashes ahead with the swiftness that comes not from speed but from a superb power of distillation. Everything is lean, yet everything is rich. This we expect from Bergman. What might not have been expected, and what is highly gratifying, is that he has found an answer in art to what lately has been troubling his art.

In his last three serious films—*Through a Glass Darkly, Winter Light,* and *The Silence*—Bergman has used increasingly parsimonious means for increasingly subjective exploration. (I omit his recent comedy, *All These Women,* and I wish *he* had omitted it. Bergman has little wit or humor except peripherally when treating a serious subject.) These films were masterfully made, but they seemed introspectively remote rather than dramatized, so much so that they gave the viewer almost a sense of intrusion. I had the growing fear that Bergman, his breathtaking techniques undiminished, his power with actors as full as ever, had become disheartened: by a sense of irrelevance, *his* irrelevance; by the imperative to choose what to communicate, by the hopelessness of choosing, by the hopelessness of finding artistic means after he had chosen. It seemed as if, in refuge, he was keeping a kind of private journal in public. But *Persona* is a successful work of art, and what is especially happy about it is that Bergman, far from abandoning the psychical questions that consume him, has plunged further into them. He has made his film unfold its matter *at* us, instead of hugging it close.

Most of *Persona* consists of the isolation of these two young women, patient and nurse, in the cottage by the sea: the mute actress, the talkative nurse. The latter is lively, frank, and intelligent, without being in the least intellectual—a girl who has reasoned out her attitude toward life. She has had her heartbreak: a long love affair with a married man that came to nothing. She has had her shock at learning what she is capable of doing: a casual but wild sex orgy (which she recounts) that she found herself enjoying. She is now engaged to a man whom she wants to marry but whom she does not love. In short, she has had some experience of disappointment and of self-surprise and is still capable of engaging life—she *wants* to engage it.

The drama consists of the process through which, in their isolation and through the nurse's virtually solo discourse, she almost seduces herself into the actress's condition, almost talks and frets herself into psychological identity with her patient. There is, to begin with, a distinct physical resemblance between them. (Bergman has said that this is what started him thinking about this film.) There are other points of contact: sufficient disgust in the nurse with herself and the world, sufficient admiration in her—almost schoolgirlish—

for the Artist. We have a growing sense that the actress is abetting the process. Through her very silence she is "using" the nurse, partly out of genuine interest, partly out of need for entertainment, but partly out of diabolic tease, the hatred of the sick for the well—or at least of the hampered for the relatively unhampered. The teasing culminates when the actress deliberately leaves unsealed a letter to be mailed by the nurse, in which she speaks patronizingly of her companion.

This unsealed letter, which is of course read by the nurse, is the turning point of the film, of the nurse's mental state. She ceases to be a nurse; she becomes an antagonist, a competitor. Her ministering and her adulation are superseded by a desire to strike back, even by an (unconscious) desire to join the actress in the shadow world—there to beat her, meld with her, somehow be *acknowledged* by her. One particularly fine sequence catapults the nurse into the realm of different realities. Barefoot and in a bathing suit, she drops a glass on the stone terrace and quickly sweeps up the pieces. Then, just as she picks up the last splinter, she hears the actress coming; she replaces the splinter on the ground and goes inside. Out comes the actress, also barefoot, and starts pacing up and down. The camera is inside with the nurse as she listens. The sharp "ouch" comes at last. Fiercely the nurse turns to the window and watches. Then the screen itself spilts, and there is a quick reprise of one of the "pre-title" shots—a nail being driven through a hand. Now the nurse is so close to the actress that she shares one of her mental images.

She is so close that later she tells the actress the reasons why the latter bore and then rejected her child. (And Bergman has her tell it twice—once as we watch the listening actress, once as we watch the nurse. The repetition has the effect of an incantation, almost a wedding ceremony.) The nurse is so close that, near the end, she herself is, for a time, reduced to speaking gibberish.

But the possible synthesis of the pair ends when the nurse, in frenzy, scratches her own wrist, and the actress—almost arrogantly, triumphantly—sucks the blood. The shock of both actions brings the nurse back to *her* reality, forces her to realize what has happened, that the nurse-patient relation has been destroyed. Then they both know that they must part. We do not actually see the actress leave, although we see her packing. But we see the nurse leave—the camera watches her past a female ship's-figurehead in the foreground. There is a split-second reference to the performance of *Electra* in which the actress first froze. The two images tell us that the nurse is figuratively leaving the actress where she found her. The nurse walks down the road and takes a bus. Then, as at the start, we see a motion picture projector; the reel of film unravels, the arc light fizzles out, the film of the film finishes. This is no simplistic sign that we have been watching illusion. Bergman is no surer than anyone else (he seems to say) as to what illusion is. He is giving us a metaphor of the film's ambiguous subject, sanity and madness: a realm which, as Laing tells us, is self-defined, not platonically absolute. Bergman's drama is in the attraction of the truth of the "true" inner self (Laing's term) as against the generally prevailing and venerated falsity of the outer world. At the last the nurse pulls free of the actress's state, not because of any indisputable and superior standard of rationality but because of her own irrationality. That, I think, is the essence of the film. If we talk of reason, there is probably as much reason on the mute actress's side, on the side of withdrawal, of inner purgation. What moves the nurse finally is a stubborn *ir*rational will to live— to live in the majority's terms, in terms of the world's continuity.

Artistically this film succeeds because its action and its symbols are more than recognizable, they are disturbing. All of Bergman's mature films are miracles of technique, but the recent ones have seemed picture exhibitions of familiar concepts of our era, neatly stated. *Persona* does not break fresh ground (hardly a requisite of art), but it throws a hot light on certain ideas that make them more painful than ever. The actress's state is so compelling, the nurse's desire to join her is so touching, that we are lashed to this film as to our own psyches (and to our own unacknowledged longings for withdrawal). The bitterness of the "healthy" ending makes this all the more true.

The visual quality is exquisite. (Sven Nykvist is again Bergman's magician-photographer.) This quality gives the needed ambiguity to the dream visit of the actress's persumably blind husband when he makes love to the nurse, thinking it is his wife, with the actress standing by. It converts what may be a lesbian encounter from the literally sexual into the chemically unified. The shot of the two women, wearing broad-brimmed hats in the sun, peeling mushrooms and humming; a long shot of the nurse standing by a pond on a rainy day, or the sequence in which she sits through a night alone on the beach; the scene in which the light deepens on the actress's face in her hospital bed as the radio plays Bach—visual experiences such as these contain the film in little and may also be what actively survives of it in memory.

Imperfections in *Persona* are few and therefore obtrusive. The moment—after the twice-told tale of the actress's son—when half of one woman's face is matched with half of the other's is heavy, superfluous. A photograph of German soldiers and Jewish civilians falls quite fortuitously out of a book to remind the actress of the world's horror. The film is so fine that the fingers itch to tear out these and a few other blemishes.

Liv Ullmann, a Norwegian, makes her first appearance in America in the role of the actress. (She has since made another film with Bergman.) Because she speaks four lines at most, it is too early to form a complete judgment of her, but she has a wonderfully taking, intelligent face, and she has the ability to listen evocatively. Bibi Andersson, one of the best young actresses on the screen, plays the nurse with such subtle yet straightforward truth that she quite erases the effect of her recent excursions into American nonsense (*Duel at Diablo*) and Swedish nonsense (*My Sister, My Love*).

Bonnie and Clyde

(*New American Review #2, January 1968*)

LITERATURE has been the source of content for innumerable films, but in *Bonnie and Clyde* it is the source of form. Such literary influence is supposed by some to be the enemy of the film medium, yet it has helped Arthur Penn to make his best picture.

That is qualified praise. His past work includes *The Chase* (horrendous), *Mickey One* (pretentious), *The Miracle Worker* (imitative), and *The Left-Handed Gun* (affected). *Bonnie and Clyde* adds up to much less than the sum of its parts, but some of its parts are extraordinarily good, and one of those parts is its structure.

It is built on two interlocking views of a literary form: the ballad. The story is based, loosely, on the lives of Bonnie Parker and Clyde Barrow, the killer-bandits of Texas and the Midwest in the early 1930s who were finally gunned down after a number of robberies and murders. The thematic apex of the film is a ballad that Bonnie writes and publishes in a newspaper (this is factual) about her and Clyde's exploits. But Penn has also conceived of the entire film as a ballad. Many sequences end with a flourish of twangy banjo music, as if a stanza had just concluded. The attitude of the film toward its hero and heroine—outlawed, brave, doomed—and the subject matter itself are those of the traditional ballad, which, as Lionel Trilling has noted, usually concerns an act of violence.

From these interlocking views come the film's virtues, which are all the result of a hard look at naturalism with romantic eyes. Penn treats killing as killing; his actors don't just fall down, they die. Still he understands the romantic connection between outlawry and space. Gangsters operate in cities, but gangsterism is never romantic (though it is often sexually exciting). The outlaw lives in open country—the West, Sicily, Australia, the Scottish Highlands. The bravado of sortie and escape, the association of nobility with open space and ignobility with the confines of grubbers, these attitudes underlie the film. Burnett Guffey's color camerawork articulates these feelings; he combines the starkness of East Texas and the Midwest with a haunting, almost picturesque quality. (In one lucky moment a cloud crosses the sun during a long shot of Clyde chasing Bonnie through a cornfield.) The 1930s clothes, designed by Theadora Van Runkle, are as veristic and nostalgic as old phonograph records. Dede Allen's editing gives a long melancholy sweep to some sequences but also gives a terrible flesh-thumping impact to the gun fights.

The film has other assets. Its chief performances are in tune with its double view. Warren Beatty, who is also the producer, plays Clyde and at last gets past his sales image of the mumbling small-town rooster into admirable theatricality and color. Faye Dunaway, as Bonnie, has the right scrawny itch of the café waitress who first becomes aware of frustrations through her genitals but discovers she is more complex than she knew. Gene Hackman, as Clyde's brother and accomplice (often sounding and looking like a younger LBJ), makes a vivid, homespun killer. Estelle Parsons, as his wife, clucks like a hen among hawks, a plump bourgeoise who, by the accident of marriage, finds herself an outlaw. And Michael J. Pollard, one of the gang, is affecting as Aaron Slick transported from Punkin Crick and given a submachine gun. All of them owe a good deal to Penn's ability to direct actors.

But the film owes its ultimate failure, I think, to Penn's basic shortcomings. The first screenplay by David Newman and Robert Benton was, reportedly, written before Penn entered the project; but the final result seems to show his influence—at least it reflects Penn's profile in the way the facts of the story have been altered. Alteration, as such, is of course perfectly in order: for instance, if Penn did not want an ugly Bonnie, a Bonnie who was married when she met Clyde and who continued to have a busy sex life of an eccentric kind with him and many others, the alteration of these facts could be ascribed to dramatic license. But most of the changes and emphases have been devised for a particular kind of aggrandizement —a ring of contemporary resonance that is hollow.

First, there is the Freudian theme. When Clyde and Bonnie meet, he proves to her that he is dangerous (virile) by showing her his revolver; she then fingers the barrel. When she becomes violently excited by watching him hold up a store, he confesses that he is impotent. She makes various attempts through the film to rouse him (including the clearest suggestion of fellation that I have so far seen in an American picture—France, look to thy laurels), but he remains inactive. It is only when she reads him the published ballad of their career, near the finish, that he is able to function, and it is only then that we understand why the subject of impotence has been included. When he feels that he *is* somebody, has been immortalized, then he can be male. This sex-and-selfhood complex is an invention of the scriptwriters, which would not matter except that they cannot make it ring true. They have to make Clyde explain endlessly why he deliberately chose this pretty girl as a partner (when he first sees her, she is naked!), knowing that it must lead to his chagrin. The real explanation for this Freudian theme, I think, is that it is part of the entertainment industry's new intellectual veneer.

This is even more true of the other dominant theme, economic determinism, which is patently fabricated. We are to believe that Clyde becomes a robber of banks because, in the Depression, he comes to identify banks with oppression; yet he committed armed robbery, and was imprisoned for it, before he makes this identification. Policemen, to the film's Clyde, are quasi-autonomous enemies of the poor; the scriptwriters keep him from seeing that policemen are usually poor men, too, and that—in those Depression days—they felt very lucky to have jobs and would do almost anything to keep them. (So would any of those dispossessed farmers, if they had been able to get police jobs.) Clyde never kills anyone until a grocer whom he is robbing tries to kill *him*. This shocks Clyde—the discovery that people will kill to protect their property. Yet he carried a pistol long before that and would presumably have used it if necessary. Besides, all he himself does in the film is acquire property, as far as he is able; he is no Robin Hood. There is a scene in an Okie camp (beautifully photographed) in which the dispossessed farmers treat the wounded bandits like People's Heroes, but there is next to no basis in the film for this attitude.

Both themes, proletarian and Freudian, are hollow, purely assumed. In the film's *own terms,* nothing more than thrill-seeking and self-aggrandizement starts the pair on their career. The scenes meant to create sympathy for them—like the horseplay between the Barrow brothers, and the Parker family reunion—are blatantly laid on, like scenes in the thirties propaganda plays that showed the embattled worker with a happy, if harried, home life. As a result, when the police finally ambush the pair and splatter them with bullets, when Clyde falls to the ground in "poetic" slow motion and the already dead Bonnie jiggles on the car seat under the continuing fusillade, we feel little more than that this is the way the picture ends. There is no horror at the grimness of society's grinding, none of the noble arch of the folk epic that was presumably intended. Our sympathy has been dissipated by the dozens of other victims of society— who happened to be tellers or policemen—already killed by the Barrow gang. Bonnie and Clyde simply go over the cliff they have been headed toward all along; there is no tragedy.

This disappointing picture is a superior example of an inferior breed: the film of make-believe meaning. Changes in America have, inevitably, changed the tone of its film industry; a college-bred generation of producers and directors (and screenwriters and publicity men) has come into being—quite different in self-estimate and status hunger from the first few generations of American film workers. (On my last trip to Hollywood I visited the set of a machine-made situation comedy; when a break was called, the thick-spectacled young assistant director put down the script and picked up a collection of the short novels of Henry James.) This latest film-making generation that has come to power (to *power*—as opposed to small independent or "underground" film makers) operates comfortably within a cosmos of intense commercial pressure to which these men have nicely adjusted their ambitions for intellectual prestige. But this reconciliation prevents them from making the sheer entertainments, comic or serious, of the palmy Hollywood days—the "sincere" days, as Jean-Luc Godard once described them with peculiar accuracy; and of course it also prevents fidelity to art and intellect. What we get are entertainment films on which "meaning" is either grossly impasted or is clung to only as long as convenient. For instance, the film of *Up the Down Staircase* takes several of the harshest problems of urban education and faces them with new contemporary honesty— until it turns its back on them.

It is relevant, I think, that many of this new generation of American directors—Penn, Sidney Lumet, Robert Mulligan, John Frankenheimer, Stuart Rosenberg, Irvin Kershner, Norman Jewison—come from television, and most of them from television's so-called golden age: that is, the postwar decade when TV was mouthing some serious ideas in half-hour or one-hour or hour-and-a-half segments, precisely measured, with interstices for commercials and with no offense to sponsors. These men are thus adept in idea-tailoring, either cutting serious material to measure or embroidering lesser material with seriousness. All of them have directorial skill; some of them have a great deal of it. What they do not have is wholeness of being, expressed wholly in their films—the wholeness that distinguishes a range of foreign directors, from the best to the good, from Bergman and Antonioni to Alain Jessua and Ermanno Olmi.

Visually these American directors are acute, and they have helped to make fine photography a commonplace in American films. But even visually they betray themselves. They try to give weight to flimsy material with superb camerawork (Haskell Wexler's superb photography for a gimmicky race-relations thriller, *In the Heat of the Night,* directed by Jewison). They use close-ups that are meant to seem unconventionally truthful but that dare nothing and say nothing (a dead dog's paw, a singing convict's mouth in Rosenberg's *Cool Hand Luke*). They strain to include entire sequences that are only inserted Arias for Cinematographer (the Parker family reunion in *Bonnie and Clyde*). Pictorially as well as intellectually, they are clever utilizers.

These new directors know enough about art and ideas to feed the ravenous appetite of the new middle class for culture status, and still not permit art and ideas to get out of hand—to have any of the results for which they were devised. The Western becomes adult and the crime film becomes Freudo-Marxist so that we can go to Westerns and crime films without skulking in and out of the theater. The glossy marital comedy (Stanley Donen's *Two for the Road*) pilfers enough from new French film art so that we can know we are "keeping up" as well as enjoying ourselves. (It even gets praised for this pilfering as proof that the commercial film is maturing. Some years ago Lucky Strike based a magazine ad campaign on Mondrian themes and was praised for maturing the ad business.) The film of make-believe meaning makes everybody feel smarter without risking anything.

None of this is to whip further the well-whipped middle class. That would be like beating a live horse—which is already galloping. It is simply to identify a contemporary phenomenon. A generation ago, directors resembling the ones I have mentioned might have been called sellouts. These men today are not sellouts. They betray nothing in themselves. They have been educated and conditioned by their culture to serve their culture, which they do without unease and with much finesse.

Perhaps we are in a transitional state as a result of the national surge in education, en route to a genuinely demanding public. We can certainly hope so; we cannot yet believe so. Meanwhile, though these directors probably feel "sincere" in Godard's sense, to an observer they seem laden with pretense. I prefer *Little Caesar* and *The Public Enemy* to *Bonnie and Clyde;* to me they are "sincere" pictures, uninhibited examples of popular art. On the other hand I also prefer Francesco Rosi's *Salvatore Giuliano* to *Bonnie and Clyde*. Rosi's Marxist film (about the Sicilian bandit) is faultily built, but it is unmistakably Rosi's complete response to the story of Giuliano, and it was made for no other reason on earth but to express the wholeness of Rosi's being on that subject. It is not a star vehicle nor a culture comfort-station for (better) book club members. Neither of these denials applies to *Bonnie and Clyde*.

POSTSCRIPT. The above was written in the summer of 1967. Within two and a half years, there had been a transition to at least one new kind of "genuinely demanding public." The baccalaureate bourgeoisie had been strenuously jostled by the anti-bourgeois young. (Further discussion of this later.) The fat-cat sureness of fortyish Hollywood moguls has been badly shaken now, as previously dependable patterns of film making drew small audiences and as other films were unpredictably successful. Worse for the moguls, there weren't even any new patterns to be deduced from the new successes. I happened to attend a trade screening of *Midnight Cowboy* shortly before it opened in New York, which was well after the epochal success of *The Graduate* and the solid success of *Bonnie and Clyde*. That screening room held about sixty film business people and theater owners, and when the picture was finished, there was silence; then there were a lot of uneasy little sideways smiles and a growing atmosphere of quiet courage in the face of what looked like certain financial disaster.

The success of *Bonnie and Clyde*, in relation to the young audience, was somewhat different from that of *The Graduate*. The latter obviously dealt directly with them, the former was taken analogically. A lot of different admirations that I heard for Penn's picture at colleges and universities around the country could be summed up in one reference that was made to a maxim of Chairman Mao's: in a time of injustice, the honest course is to be a bandit. The analogy seemed, and seems, superficial. The two killer-lovers of this film are adventurers with strong sex-ego drives, with only patches of social consciousness patently stitched onto them. They are a long way, for instance, from the young Jean Genêt, who lived by theft in France but refused to steal in Nazi Germany because it provided too little contrast with the national environment.

There is no point in arguing with the fact of *Bonnie and Clyde*'s success; nor have I tried above to diminish what I think are its virtues; but the superstructure of ideas built up around it by some young people seems to me to blur and to magnify a saga of egocentric anarchy.

Ulysses; Passages from James Joyce's Finnegans Wake

(New American Review #2, January 1968)

SUBJECTIVITY—intense exploration of persona and person—has fascinated film makers for decades, so I suppose it was inevitable that the two greatest novels of subjectivity in English—*Ulysses* and *Finnegans Wake*—would someday reach the screen. The decisive difference between the two films is that one of them recognizes the impossibility of its task and has considerable merit, the other seems to believe that it has done the job and is a considerable dud.

The dud is *Ulysses,* adapted by Joseph Strick and Fred Haines and directed by Strick. Because the film's assumption is that it has encompassed the book, we must start with what the book is about. There is little mystery on that point. Joyce's general intent has been well described by Alfred Kazin: "He sought to bring the largest possible quantity of human life under the discipline of the observing mind, and the mark of his success is that he gave an epic form to what remains invisible to most novelists." Specifically for *Ulysses,* one can justly (I believe) conjecture Joyce's initial premise: If I take an average day in the life of an average man and tell *everything* about it—in act and thought and half-thought and association—I will describe the history and condition of contemporary Western man.

What do we get from Strick? A film that consists mainly of, on the one hand, pleas for brotherhood—anti-anti-Semitism—and, on the other hand, sexual candor. Candor in word, not in act. These elements are of course in the novel, but they are only parts of the novel and are quite carefully integrated. The novel's anti-anti-Semitism is not fed us in Stanley Kramer spoonfuls. In fact it is not brotherhood propaganda at all: Leopold Bloom's status as a Jew in Dublin symbolizes every man's status as an outsider in modern society. The sexual frankness is hardly a minor part of the book; Joyce was one of the first novelists to insist on the omnipresence of sexual drives. One can even see him "trying out" sexual frankness in his early letters to his wife (in those of his letters, one may add, that can be published). But to wrench sex out of the general "invisible" texture of *Ulysses* is sheer vandalism.

Many people have been impressed because there is nothing in

Strick's film that is not in the book.* This is a good Hollywood definition of integrity. It does not bother these people that there is far *less*
in the film than in the book: that to reproduce an ear and a leg does
not reproduce a whole man, no matter how faithfully the ear and leg
are reproduced. To insist that ear-and-leg fidelity is fidelity to the
whole man is phony praise—particularly when there is a very good
market for ears and legs at the moment.

Concede the necessity to shoot the film in present-day Dublin, even
though some of the dialogue and attitudes sound odd in a 1960s
setting. Concede that such a passage as the birth of Mina Purefoy's
baby could not be reproduced on film because it depends on a progression of prose styles for which there is no cinematic analogue. Do
not even object because all the mythological references are omitted—
so that it is impossible from this film to understand the title. Still we
can ask why the things that the film has attempted have not been done
well. There is some good, brisk, youthful feeling in the opening at
the Martello tower; past that, the going is lame. Outstandingly inept
is the *Walpurgisnacht* scene in the Dublin stews, which sinks to the
imaginatively limp level of Strick's film made from Genêt's brothel
play, *The Balcony.* Surely if there is one power easily in the film's
grasp, it is fluidity—the ability to suggest flowing dream states. Strick
gives us a series of photographs—head-on, cornily lighted, edited
in clickety-clack style. Molly Bloom's codal soliloquy is simply a
monologue with illustrations, with nothing like the torrential effect
that it needs. And, of course, the soliloquy is edited to concentrate on
the sexual passages. This is "fearlessness" and "honesty."

Barbara Jefford, as Molly, is the most competent of the cast, though
not my image of that domineering, pathetic, vaginal universe of a
woman. Maurice Roeves, the Dedalus, is too handsome for the part.
Milo O'Shea, the Bloom, is an empty moonfaced vaudevillian,
utterly at sea. This would be clear in any event; it is doubly clear after
having seen Zero Mostel (tricky though he was) in the role on stage.

In contrast with Strick's effort, Mary Ellen Bute's film is called
Passages from James Joyce's Finnegans Wake, and with this modest
claim, it succeeds surprisingly well. Miss Bute had to deal with less
story as such than there is in *Ulysses,* but she also had to decide on all

* Richard Ellmann points out that this is not quite true, that there are "rare
and infelicitous interpolations." (*New York Review,* June 15, 1967)

settings and in many cases to select the speakers. She based her film script on Mary Manning's dramatization, in regard to which Denis Johnston wrote: "Joyce . . . presents us with a play where, as in the Song of Solomon, we are expected to work out for ourselves who it is that is reporting or being reported in every line." Even more stringently than the stage adapter, the film maker is pressed against the thorns of this problem every split second of the time. In the main Miss Bute has coped with it sympathetically and impressively. The proof is not in any checking that can be done—one opinion is as good as another—but that, by and large, the film achieves the innermost effect of the great dream novel.

It is a bit too long: ten minutes less (it runs 97 minutes) would be twenty minutes better. Some of it is as cornily lighted as Strick's *Ulysses.* The music is often sheer movie music. Some of the pictorial compositions have a whiff of home movies; others have a whiff of the film school. But what keeps the film from slipping into amateurism, what redeems it far past its faults, is the very strong sense of Miss Bute's vision: her loving perception of the novel and her response to it in cinematic imagination. She has imposed a stern dream logic on her material. She extracts a sequence, almost arbitrarily, from the novel; just when it shows signs of disintegrating, she cuts to a new sequence, which not only starts well but has the effect of reaffirming the sequence we have just left, because the rigor of the change reaffirms Miss Bute's grip of the work. This is doubly so because, dramatically and pictorially, Miss Bute has had virtually to invent every sequence.

Subtitles are used. Nearly every word that the actors speak is flashed on the screen. Ordinarily this would be deplorable (except with foreign-language films, where it is still the best method devised for making them available to us). But *Finnegans Wake* is not an ordinary case, and here the practice is excellent. First, the composite words are impossible to understand through the ear alone. Second, these Joycean words are wonderful visual *objects,* enriching the picture *visually,* like objects in surrealistic art.

The cast, mostly Irish, speak the composite language with such conviction that the reality of a million compressed dreams makes our heads appropriately swim. Martin J. Kelley as Finnegan etcetera, and Jane Reilly as Anna Livia etcetera are always effective. It is impossible to say that they play their roles well because there are no roles, but what they have to do, they do interestingly.

The film's central achievement is that it touches myth, touches our old friend the collective unconscious. I heard someone say after seeing Strick's *Ulysses* that it would make a good introduction to the novel. The remark made my blood run cold because it seems both inevitable and dreadful: the film is a facile and ludicrous reduction. But Miss Bute's film, modest and flawed as it is, is in tune with the work it is about, and can even be seen as a small introductory suite to the teeming Joycean opera.

Cool Hand Luke

(*New American Review #2, January 1968*)

Cool Hand Luke, in which Paul Newman gives still another performance so easily persuasive that he makes good acting look easy, is about a Southern chain gang; and it contrasts revealingly with the celebrated *I Am a Fugitive from a Chain Gang* (1932). In the latter, Paul Muni played an innocent man unjustly convicted, who struggled against the horrors of the prison camp, was goaded to escape, and was at last condemned by society to be the criminal he had not been. Newman's character is guilty from the start—and guilty of wanton silly destruction (destroying parking meters when drunk), not any kind of purposeful crime. He is not especially outraged by the harsh camp conditions. Unlike Muni's character, he has no goal of social reacceptance that he is fighting for. His escapes are self-willed escapades, not acts of heroism. And he gets himself killed out of stubborn cussedness, not for any cause or any practical reason. Thus the popular film arrives in the age of anti-idealism and of the *acte gratuit.* Camus's Absurdity on the quarter shell.

Closely Watched Trains; How I Won the War

(New American Review #2, January 1968)

Two comments on two important films. Both come from abroad and both are about the Second World War.

Closely Watched Trains, a Czech film made in 1966, was directed by Jiri Menzel, who, like most young East European film directors, is a graduate of his country's film school; like many good directors outside the U.S., he is coauthor of his screenplay; like a surprising number of young foreign directors, he is also an actor. (He plays a small part in this film.) Menzel, born in 1938, is too young to have had much immediate reaction to the German occupation of Czechoslovakia, so his film, written with Bohumil Hrabal, is a view of immediately inherited history, rather than of experience. This combination of closeness and distance may be what gives him his cool style without loss of central compassion.

The film is set in a rural railroad station. His protagonist is a late adolescent who goes to work in that station. The boy, played delicately by Vaclav Neckar, is shy, eager, touchingly dignified. A pretty round-cheeked girl who works as a conductor is fond of him; a masterful philanderer who works in the station uncles him; the stationmaster benevolently tyrannizes him. He is grateful for the affection, the patronization, the discipline. The film is a small *Entwicklungsroman* of his passage from his mother's sheltering home into the world.

If I seem to have switched themes, that effect is the quintessence of Menzel's view. He concentrates fiercely on the daily life of this boy, his job, his ambitions, his unsuccessful first bedding with his girl, his worries about his manhood. It is only obliquely that the Nazi occupation enters the film and never, really, until the end does it come full center. Menzel is acknowledging that boys have ambitions, get erections, emulate their elders, indulge in daydreams, no matter what chief sits in the capital. He is also saying that, in an ancient country, there is an ancient schism between the peasantry and the government, whether the government is monarchical, fascist, democratic, or socialist. The peasant's first duty is to survive, despite the efforts of government to hinder or help him. And the facts of war—of whatever particular war it happens to be at the moment—are simply one more condition in his struggle.

This theme has been treated before, notably in *Two Women,* the film by De Sica and Zavattini out of Moravia, but there a somewhat broader brush was used for a more traditional humanitarian approach. Menzel, much younger than the Italian partners, more tart and elliptical, focuses more thoroughly on the minutiae of daily existence, edits more brusquely, and indeed sees the whole grim era from a wry, sardonic angle. (The only inconsistent note of movie contrivance is that, on the night before the youth is killed, Menzel allows him his first full sexual experience—with the female member of the resistance who has unwittingly brought him his death warrant.) The boy goes to his death without any tinge of heroism; he is simply doing his neighborly peasant duty in the resistance as he did his duty in his railway job. Typical of the film's understatement is the moment of his death on the signal tower. We don't see him clutch himself as he is shot, no spasms.

We hear the shot, then we see him sprawled on top of a freight car passing beneath the tower, being borne away. Everything gets borne away sooner or later, the film seems to say; the question is: Is this a sufficient reason not to care? By the very selection of his theme, Menzel seems to answer in the negative.

Closely Watched Trains, a far superior work to the recent Czech successes *The Shop on Main Street* and *Loves of a Blonde,* is the best film I have seen from the new wave of Czech film making.

How I Won the War is a British film directed by an American, Richard Lester. Lester's two early features, the Beatles films, were happy surprises (particularly the first). It is rare that popular singers make a film that is something other than a crudely hemstitched series of singing acts. Lester "read" the Beatles' characters and rendered them in freehand sketches. His version of *The Knack* was one of the most completely cinematic translations of a play that I know. But *A Funny Thing Happened on the Way to the Forum* was a disaster because a vehicle for some utterly theatrical stars ran headlong into a director for whom the camera is the star. Still, even the least of these films had scintillating moments, and the best of them are fireworks displays that spell out some secrets of our times. His new film is Lester at his very best.

The script of *How I Won the War* was adapted by Charles Wood from a novel by Patrick Ryan with even busier disregard for the original than Wood showed for the original of *The Knack,* which he also adapted. He has given Lester the requisite freedom to play— to play, I repeat—with the subject of World War II. The story deals with a goodhearted, muttonheaded young lieutenant who leads his troop through the war to their destruction. Only two survive—himself and the troop coward—yet, as he says with figurative truth, he won the war.

The film begins in early 1945 at the Rhine, where the lieutenant is captured, and most of the film is flashback, from training days through the campaign in North Africa. But that is too formal a description; there are no orthodox flashbacks; there is only an unremitting series of flashes—back, forward to the "present," further forward to the future, into fantasy and extrapolation. There are two main story elements: in North Africa the troop is sent behind the German lines to prepare a cricket pitch so that an English general will be able to play after he advances; at the Rhine the lieutenant, in German hands, "buys" the Remagen bridge from a German officer so that the Allies can speed ahead. These two elements are not funny, they are ridiculous. They are thus perfect for Lester's purpose.

The film's first few moments are discouraging, as if Lester had sunk from *Forum* to *Carry On, Lieutenant,* and throughout there are additional jarring lapses. For instance, the lieutenant is thrown headfirst into sand, and the men gloat at his muffled cries for help. But most of the script, filigreed with good wiry dialogue, serves as a fine trampoline for Lester. These are numerous "plants" of material—visual and verbal—to which later reference is made, too neatly modulated to be called running gags. There is a barrage of parodic transformations. For instance, a blimpish colonel gives the lieutenant a gung-ho speech in a dugout. When the camera pulls back at the end of his exhortation, the dugout—suddenly—is on a stage, and the curtain descends as the colonel finishes roundly. (Lester does not leave it there. The audience in that theater is sparse and the applause is slack.) A number of incidents are swiftly replayed in different settings, as in a spoof of *Marienbad.* The music yawns scoffingly: whenever we cut back to these bedraggled desert rats, we get a swell of grandiose Oriental goo on the soundtrack in *Lawrence of Arabia* style. And we are continually reminded that the whole thing is a film. When one of the men is hysterical, another soldier turns to the audience and says angrily, "Would you be so kind as to take that camera away?" and we flash to a shot of two cockney biddies in a cinema watching the awful scene comfily. At the end, as the war is finishing, two soldiers discuss what they are going to do next and think they may get work in a film that is going to be made about Vietnam. (There is a marked difference here from the "film-consciousness" of *Persona*. In the latter, Bergman reminds us that we and he are involved in a film. But Lester tells us that we, he, *and the actors* know that it's a film.)

Most of this film is in color, but there are many black-and-white sequences. The latter often look like newsreels, but Lester says that only eighty-two feet of the film are newsreel clips. He shot the troop's ludicrous exploits in color and the context of "real" war in black-and-white (sometimes tinted); as the troop moves into the "real" war, they also move into black-and-white. The conclusion, again ludicrous, is again in color.

The dominant note is sounded in one flintily funny device. Whenever a member of the troop is killed, he reappears, shortly after, clad entirely in one color, and he keeps going with the group. The first dead man comes back in green—entirely green, including a green silk stocking over his face. He simply appears, is paid no special notice by the others, and carries on with his duties. We stiffen with apprehension at some heavy symbolic Unknown Soldier significance. Then the lieutenant addresses his men, including the green one, strides past them, turns back to the green man, and says, "Do you think you ought to report sick?" By the end there are pink and purple and other colored soldiers, all dead, all continuing as comic objects and objects of comedy.

In this same heartless comic way Lester is also very touching. One man is badly wounded in the legs, is left lying in the sands, and starts talking deliriously to his wife. Then she walks up the dune in her London house dress—no misty fade-in; she just walks up the slope mouthing BBC platitudes, then starts to comfort him as if he had sprained his ankle on the front step. "It hurts, Flo," he says quietly. "Run them under the cold tap, love," she says equally quietly.

The actors are so good that they seem both to give excellent performances and merely to be Lester's instruments. They include Michael Crawford, as the piping, eager lieutenant; Michael Hordern, as the impenetrable colonel; Roy Kinnear doing a comic version of the fat Tommy he played in Lumet's *The Hill;* Jack MacGowran, as a former music hall comic whose "turns" keep turning up surrealistically in the desert; John Lennon, a Beatle on leave, as a former Mosley man—all coolly insane. Lee Montague, the corporal, is hopelessly sane.

With the light-fingered help of his editor, John Victor Smith, Lester caroms the film off a series of colored bubble gum pictures of WW II battles—cheap unreal icons of events that Lester makes more truly unreal. Yes, the film tells us that war is hell and that the most hellish thing about it is that fundamentally men love it. But Lester is telling it all from a particular point of view. When Lester's film is not (occasionally) straining to be funny, it is *genuinely* not funny—the kind of comedy at which one does not laugh, comedy that seems to take place in a cavern of ice where all the laughter has already been laughed, has been caught and frozen in glittering, frightening stalactites.

This is because Lester's film speaks from the very center of the Age of the Put-on. This is the heart of the sixties speaking about Dunkirk, Alamein, the murder of the Jews. No German officer would actually have said what Lester's German says about murdering Jews. (He is quite unperturbed—neither triumphant nor vicious nor tormented.) This film is the Mod generation's view of the war and of our mourning for it—a view of history not as tragedy but as stupidity. To them, we of an older generation who are still involved with such matters as this century's evil and guilt are simply entangled, in another way, in the same stupidity. And stupidity is funny; but this stupidity was so huge that it is deeper than ha-ha funny. The film's viewpoint is morally shocking, in the most serious sense, and is seriously debatable; but it is neither immoral nor amoral, and it is brilliantly, scathingly put.

In every way Lester and his generation have turned serveral pages of the calendar. They speak film language in a mode that is impossible to those twenty or thirty years older, no matter how talented the latter may be. This may not be all to the good, but it is so. There is even a difference between Lester and some of his contemporaries in East Europe, including Menzel, who is in fact six years younger. A recent festival of Czech films in New York, as well as numerous Polish and Hungarian and Yugoslavian films that I have seen here and abroad, makes clear that young directors in those countries are hungry for new Western influences; that they use these new styles for sharp criticism of their societies, often with surprisingly direct political reference. Via the route of disappointed Marxism rather than disappointed Christianity, they seem to be reaching some acceptance of man's inevitable and utter self-dependence. But Lester's attitude is not one of disappointment; the Age of the Put-on is an age of pragmatism, cynical but adventurous. This mood has not yet been reflected in any film from Eastern Europe that I know. Menzel's WW II is not Lester's WW II, although both are different from my generation's. The young Easterners are still trying to make fresh films about life. Lester sees the very concepts of film and life as an infinite series of Chinese boxes. The Easterners are not quite sure what they believe or whether belief is still possible. Lester believes completely, in his camera.

Elvira Madigan

(December 2, 1967)

A film of Beauty is a risk forever. The Swedish director Bo Widerberg has taken that risk quite deliberately in *Elvira Madigan* and has come off fairly well; but finally his film falls short not only through the inevitable tedium—in a sense, the distraction—of the incessantly beautiful, but also because of the subject on which all this beauty is heaped.

The screenplay, by Widerberg, is based on fact, we are told. In 1889 a young cavalry lieutenant, Count Sixten Sparre, leaves his wife and children to run off with a circus tightrope dancer named Elvira Madigan. They spend some idyllic weeks in the Swedish summer countryside, hiding out because he is a deserter. A brother officer finds Sixten and tries to recall him to duty and family, without success. The lovers run out of money. Rather than return to the world and to separate lives, they agree to die. He shoots her and himself.

Since this story is well known in Sweden, Widerberg's primary job was not one of narration but of satisfying previous romantic fantasies in the minds of the Swedish audience. He wanted to eliminate the element of suspense for *all* audiences by putting a note at the beginning of the film synopsizing the facts of the story. (In America, at least, the distributor refused to oblige.) Forewarned would, to a degree, be forearmed, but even without the prefatory note, this is not a film with much narrative element; it is an exploration of a *state*—of being and feeling—which eventually dissipates. Widerberg, presumably having chosen the subject precisely for this reason, has concentrated on making the texture of that state as voluptuous as possible. His success with the texture is the success, and eventually the failure, of the film.

Virtually every shot is exquisite. Even the scenes toward the end when the literally starving lovers are scrounging on all fours for mushrooms and herbs, when Elvira gets sick, are exquisitely photographed. Jorgen Persson's color camerawork has all the golden lights of summer, in noonday fields and deep glades; his interiors are grainy with wood and cool with linen. Skies are impossible, flowers float, the fruit and cheese and wine look better than they could ever taste. The faces of the lovers, Thommy Berggren and Pia Degermark, are out of Degas, and the hotel cook, Cleo Jensen, has the good-natured, gently lascivious face of all the friends of lovers since lovers needed friends.

Like endless slices of delicious cake, the shots fall before us, one after another. But the result is inevitable: the diet is *too* rich. It cloys. We become overly conscious of the industrious application of beauty, and we wonder whether, somewhere in the Swedish summer of 1889, there wasn't one unlovely prospect.

This discomfort would probably have arisen in any case, but it is emphasized here for another reason. These two lovers are stupid. What is obvious to the viewer fairly soon dawns on these two slowly and surprisingly. What are they going to do when Sixten runs out of money—the little money that he has with him? He cannot get a job. There is a brief conversation with a workman to underscore the fact that Sixten doesn't know any trade, and besides, he is a deserter, subject to arrest. Elvira makes one attempt to earn money as a dancer, but clearly that existence would be, for him, intolerable. When his small purse is empty, the idyll will be done. This seems to surprise them, along with her apparent surprise when she learns that her parents are worried or his surprise when he learns that his wife is desperately grieved. What did they expect to happen? How did they expect to live?

It seems to me that there are two kinds of all-for-love lovers who are tragic: those who make plans to beat the world and who are frustrated (Romeo and Juliet) and those who know they are doomed and go ahead anyway because they prefer doom to separation (the pair in *Mayerling*). But this pair have neither any plan nor any sense of what they are getting into. They are just dumb. Their fate has some pathos because they *have* been happy and they do end up dead, but our impatience with them spoils the intended tragic fall. All the breathtaking long shots across quiet meadows, the prospects through waving grass, seem wasted on two ninnies.

None of this is the fault of the actors. Berggren and Miss Degermark are subtle and true. Widerberg has handled them discreetly, and he has edited his film surely. Technically, his only lapse is his use of music. He plonks down a chunk of the slow movement of Mozart's 21st piano concerto whenever he wants a little more poignancy, beginning and ending the quotations abruptly. That music was lovely before Widerberg was born. It would be more of an achievement if he had commissioned the right composer, approved the right result, and mixed his sound track more gently.

The Graduate

(December 23, 1967)

HAPPY news. Mike Nichols' second film, *The Graduate,* proves that he is a genuine film director—one to be admired and concerned about. It also marks the screen debut, in the title role, of Dustin Hoffman, a young actor already known in the theater as an exceptional talent, who here increases his reputation. Also, after many months of prattle about the "new" American film (mostly occasioned by the overrated *Bonnie and Clyde*), *The Graduate* gives some substance to the contention that American films are coming of age—of our age.

The screenplay, based on a novel by Charles Webb, was written by Calder Willingham and Buck Henry. The latter, like Nichols, is an experienced satiric performer. (Henry appears in this picture as a hotel clerk.) The dialogue is sharp, hip without rupturing itself in the effort, often moving, and frequently funny except for a few obtrusive gag lines. The story is about a young cop-out who—for well-dramatized reasons—cops at least partially in again.

Benjamin is a bright college graduate who returns to his wealthy parents' Hollywood home and flops—on his bed, on the rubber raft in the pool. Politely and dispassionately, he declines the options thrust at him by barbecue pit society. The bored wife of his father's law partner seduces him. Benjamin is increasingly uncomfortable in the continuing affair, for moral reasons of an unpuritanical kind. (The bedroom scene in which Benjamin tries to get her to *talk* to him is a jewel.) The woman's daughter comes home from college, and against the mother's wishes, Benjamin takes her out. He falls in love with the girl—which is predictable but entirely credible. He is blackmailed into telling her about his affair with her mother and, in revulsion, the girl flees—back to Berkeley. Benjamin follows, hangs about the campus, almost gets her to marry him, loses her (through her father's interference), pursues her, and finally gets her. For once, a happy ending makes *us* feel happy.

To dispose at once of the tedious subject of frankness, I note that some of the language and bedroom details push that frontier (in American films) considerably ahead, but it is all so appropriate that it never has the slightest smack of daring, let alone opportunism. What is truly daring, and therefore refreshing, is the film's moral stance. Its acceptance of the fact that a young man might have an affair with a woman and still marry her daughter (a situation not exactly unheard of in America although not previously seen in American films) is part of the film's fundamental insistence: that life, today, in our world, is not worth living unless one can *prove* it day by day, by values that ring true day by day. Moral attitudes, far from relaxing, are getting stricter and stricter, and many of the shoddy moralistic acceptances that dictated mindless actions for decades are now being fiercely questioned. Benjamin is neither a laggard nor a lecher; he is, in the healthiest sense, a moralist—he wants to know the value of what he is doing. He does not rush into the affair with the mother out of any social rote of "scoring" any more than he avoids the daughter—because he has slept with her mother—out of any social rote of taboo. (In fact, although he is male and eventually succumbs, he sees the older woman's advances as a syndrome of a suspect society.) And the sexual dynamics of the story propels Benjamin past the sexual sphere; it forces him to assess and locate himself in *every* aspect.

Sheerly in terms of moral revolution, all this will seem pretty commonplace to readers of contemporary American fiction. But we are dealing here with an art form that, because of its inescapable broad-based appeal, follows well behind the front lines of moral exploration. In America it follows less closely than in some other countries, not because American audiences are necessarily less sophisticated than others (although they *are* less sophisticated than, say, Swedish audiences) but because the great expense of American production encourages a producer to cast the widest net possible. None of this is an apology for the film medium, it is a fact of the film's existence; one might as sensibly apologize for painting because it cannot be seen simultaneously by millions the way a film can. Thus the arrival of *The Graduate* can be viewed two ways. First, it is an index of moral change in a substantial segment of the American public, at least of an awakening of some doubts about past acceptances. Second, it is irrelevant that these changes are arriving in film a decade or two decades or a half century after the other arts, because their statement in film makes them intrinsically new and unique. If arts have textural differences and are not simply different envelopes for the same contents, then the *way* in which *The Graduate* affects us makes it quite a different work from the original novel (which I have not read) and from all the dozens of novels of moral disruption and exploration in recent years. Recently an Italian literary critic deplored to me the adulation by young people of films, saying that the "messages" they get from Bergman and Antonioni and Godard had been stated by the novel and even the drama thirty or forty years ago. I tried, unsuccessfully, to point out that this is not really true: that if art as art has any validity at all, then the film's peculiar sensory avenues were giving those "old" insights a presence they could not otherwise have.

This brings us to the central artist of this enterprise, Mike Nichols. In his first picture, *Who's Afraid of Virginia Woolf?*, Nichols was shackled by a famous play and by the two powerhouse stars of our time; but considering these handicaps, he did a creditable job, particularly with his actors. In *The Graduate,* uninhibited by the need to reproduce a Broadway hit and with freedom to select his cast, he has moved fully into film. He is perceptive, imaginative, witty; he has a shrewd eye, both for beauty and for visual comment; he knows how to compose and to juxtapose; he has an innate sense of the manifold ways in which film can be better than *he* is and therefore how good he can be *through* it—including the powers of expansion and ellipsis.

From the very first moment, Nichols sets the key. We see Benjamin's face large and absolutely alone. The camera pulls back, we see that he is in an airliner and a voice tells us that it is approaching Los Angeles; but Benjamin has been set for us as *alone*. We follow him through the air terminal, and he seems just as completely, even comfortably, isolated in the crowd as he does later, in a scuba suit at the bottom of his family's swimming pool, when he is huddling contentedly in an underwater corner while his twenty-first birthday party is being bulled along by his father up above.

Nichols understands sound. The device of overlapping is somewhat overused (beginning the dialogue of the next scene under the end of the present scene), but in general this effect adds to the dissolution of clock time, creating a more subjective time. Nichols' use of nonverbal sound (something like Antonioni's) does a good deal to fix subliminally the cultural locus. For instance, a jet plane swooshes overhead—unremarked—as the married woman first invites Benjamin into her house.

In *Virginia Woolf* I thought I saw some influence of Kurosawa; I think so again here, particularly in such sequences as Benjamin's welcome home party where the camera keeps close to Benjamin, panning with him as he weaves through the crowd, moving to another face only when Benjamin encounters it, as if Benjamin's attention controlled the camera's. As with Kurosawa, the effect is balletic; it seeks out quintessential rhythms in commonplace actions.

On the negative side, I disliked Nichols' recurrent affection for the splatter of headlights and sunspots on his lens; and his hang-up with a slightly heavy Godardian irony through objects. (The camera holds on a third-rate painting of a clown after the mistress walks out of the shot. When the girl leaves Benjamin in front of the monkey cage at the San Francisco zoo, the camera, too luckily, catches the sign on the cage: Do Not Tease.) And a couple of times Nichols puts his camera in places that merely make us aware of his cleverness in putting it there: inside the scuba helmet, inside an empty hotel room closet looking past the hangers.

But the influences I have cited (there are others) only show that Nichols is alive, hungry, properly ambitious; the defects only show that he is not yet entirely sure of himself. Together, these matters show him still feeling his way toward a whole style of his own. What is important is his extraordinary basic talent: humane, deft, exuberant. And I want to make much of his ability to direct actors, a factor generally overlooked in appraising film directors. (Some famous directors—Hitchcock, for example—can do little with actors. They get what the actor can supply on his own. Sometimes—again like Hitchcock—these directors seem not even aware of bad performances.) He has helped Anne Bancroft to a quiet, strong portrayal of the mistress, bitter and pitiful. With acuteness he has cast Elizabeth Wilson, a sensitive comedienne, as Benjamin's mother. From the very pretty Katharine Ross, Benjamin's girl, Nichols has got a performance of sweetness, dignity, and a compassion that is simply engulfing. Only William Daniels, as Ben's father, made me a bit uneasy. His WASP caricature (he did a younger version in *Two for the Road*) is already becoming a staple item.

In the leading role, Nichols had the sense and the courage to cast Dustin Hoffman, unknown (to the screen) and unhandsome. Hoffman's face in itself is a proof of change in American films; it is hard to imagine him in leading roles a decade ago. How unimportant, how *interesting* this quickly becomes, because Hoffman is one of the best actors of his generation, subtle, vital, and accurate. Certainly he is the best American film comedian since Jack Lemmon, and, as theatergoers know, he has a much wider range than Lemmon. (For instance, he was fine as a crabby, fortyish, nineteenth-century Russian clerk in Ronald Ribman's play *Journey of the Fifth Horse*.)

With tact and lovely understanding, Nichols and Hoffman and Miss Ross—all three—show us how this boy and girl fall into a new kind of love: a love based on recognition of identical loneliness of their side of a generational gap, a gap which—never mind how sillily it is often exploited in politics and pop culture—irrefutably exists. When her father is, understandably, enraged at the news of his wife's affair with his prospective son-in-law and hustles the girl off into another marriage, Benjamin's almost insane refusal to let her go is his refusal to let go of the one reality he has found in a world that otherwise exists behind a pane of glass. The cinema metaphors of the chase after the girl—the endless driving, the jumping in and out of his sports car, even his eventual running out of gas—have perhaps too much slapstick about them; they make the film rise too close to the surface of mere physicality; but at least the urgency never flags. At the wedding, when he finds it—and of course it is in an ultramodern church—there is a dubious hint of crucifixion as Benjamin flings his outspread arms against the (literal) pane of glass that separates him from life (the girl); but this is redeemed a minute later when, with the girl, he grabs a large cross, swings it savagely to stave off pursuers, then jams it through the handles of the front doors to lock the crowd in behind them.

The pair jump onto a passing bus (she in her wedding dress still) and sit in the back. The aged, uncomprehending passengers turn and stare at them. (One last reminder!—of Lester's old-folks chorus in *The Knack*.) Benjamin and his girl sit next to each other, breathing hard, not even laughing, just happy. Nothing is solved—none of the things that bother Benjamin—by the fact of their being together; but, for him, nothing would be worth solving without her. We know that, and she knows that, and all of us feel very, very good. The chase and the last-minute rescue (just after the ceremony is finished) are contrivances, but they are contrivances tending toward truth, not falsity, which may be one definition of good art.

Paul Simon has written rock songs for the film, sung by Simon and Garfunkel, and as in many rock songs, these lyrics deal easily with such matters as God, *Angst,* the "sound of silence," and social revision. But they are typical of the musical environment in which this boy and girl live.

Some elements of slickness and shininess in this wide-screen color film are disturbing. But despite them, despite the evident influences and defects, the picture bears the imprint of a man, a whole man, warts and all: which is a very different imprint from that of many of Nichols' highly praised, cagy, compromised American contemporaries. *All* the talents involved in *The Graduate* make it soar brightly above its shortcomings and, for reasons given, make it a milestone in American film history. Milestones do not guarantee that everything after them will be better; still they are ineradicable.

(*February 10, 1968*)

DOESN'T the film split in half? This is the recurrent question in a number of letters about *The Graduate*—although almost all correspondents start by saying they enjoyed it! I have now seen it again and have read the novel by Charles Webb on which it is based, and some further comment seems in order.

I like what I liked in the film even more, but now, having read the original, I can see a paradox about its shortcomings. (Many of which were noted in my review.) Besides the fact that a great deal of Webb's good dialogue is used in the screenplay, the structure of the first two-thirds of his book—until Benjamin goes to Berkeley—is more or less the structure of the film. The longest scene in the picture—the one in which Benjamin tries to get his mistress to talk to him—is taken almost intact from the book. But Mike Nichols and his screenwriters rightly sensed that the last third of the book bogged down in a series of discussions, that the novel's device for Benjamin's finding the place of Elaine's wedding was not only mechanical but visually sterile, and that in general this last third had to be both compressed and heightened. In reaction to the novel's weaknesses, they devised a conclusion that has weaknesses of its own. But there is a vast difference between weakness and compromise.

Benjamin does *not* change, in my view, from the hero of a serious comedy about a frustrated youth to the hero of a glossy romance; he changes *as Benjamin*. It is the difference between the women in his life that changes him. Being the person he is, he could not have been assured with Mrs. Robinson any more than he could have been ridiculous and uncommanding with Elaine. We can actually see the change happen—the scene with Elaine at the hamburger joint where he puts up the top of the car, closes the windows, and talks. *Talks* —for the first time in the film. Those who insist that Mrs. Robinson's Benjamin should be the same as Elaine's Benjamin are denying the effect of love—particularly its effect on Benjamin, to whom it is not only joy but escape from the nullity of his affair and the impending nullity of himself.

There is even a cinematic hint early in the picture of the change that is to come. Our first glimpse of Mrs. Robinson's nudity is a reflection in the glass covering her daughter's portrait.

In character and in moral focus the film does not split, but there is a fundamental weakness in the novel which the film tries, not quite successfully, to escape. The pivot of action shifts, after the story goes to Berkeley, from Benjamin to Elaine. From then on, he knows what he wants; it is she who has to work through an internal crisis. It was Nichols' job to dramatize this crisis without abandoning his protagonist, to show the girl adjusting to the shocking fact of Benjamin's affair with her mother, and he had to show it with, so to speak, only a series of visits by the girl to the picture. To make it worse, the environment—of the conventional campus romantic comedy—works against the seriousness of the material. The library, the quad, the college corridor have to be *overcome,* in a sense. Nichols never lets up his pressure on what he feels the film is about, but the obliqueness of the action and the associative drawbacks of the locale never quite cease to be difficulties. And, as I noted, the final chase—though well done—gets thin.

But I think that, with some viewers, Nichols also suffers for his virtues. He has played to his strength, which is comedy; with all its touching moments and its essential seriousness, this is a very funny picture. To some viewers, a comedy about a young man and his father's partner's wife immediately seems adventurous; a comedy about a young man and a girl automatically gets shoved into a pigeonhole. This latter derogation seems to me unjust. We have only to remember (and to me it is unforgettable) that what is separating these young lovers is not a broken date or a trivial quarrel but a deep taboo in our society. For me, the end proof of the picture's depth is the climax in the church, with Dustin Hoffman (even more moving the second time I saw him) screaming the girl's name from behind the glass wall. A light romance? That is a naked, last cry to the girl to free herself of the meaningless taboo, to join him in trying to find some possible new truth.

Yes, there are weaknesses. Yes, there are some really egregious gags. ("Are you looking for an affair?" the hotel clerk asks the confused Benjamin in the lobby.) But in cinematic skill, in intent, in sheer connection with us, *The Graduate* is, if I may repeat it, a milestone in American film history.

Postscript. In an interview with Joseph Gelmis (*The Film Director as Superstar,* Doubleday, 1970), Nichols apparently refers to my remarks about Kurosawa's influence on him and says that he never saw a Kurosawa film until after he finished *The Graduate. Mea culpa,* but . . .

Attribution is always risky but always tempting, particularly when one admires an artist and is trying to "place" him as part of that admiration. Nichols also says, "I've been incredibly influenced but you can't tell me by whom. I defy you to tell me by whom." I accepted the challenge in advance and was wrong. Directors other than Kurosawa have used a traveling camera to track one face to another, and presumably Nichols has seen the device before. But Kurosawa uses it with a dramatic tension that is like the order of planets in space, bound and orbited by gravitational pulls. Nichols got some of the same feeling without having seen Kurosawa's work; even more praise to him, then. (But he goes on to say that George Stevens has been "very important" to him. Stevens, according to Donald Richie, had also been important to Kurosawa!)

After the first of my two reviews appeared, Charles Webb, the author of the novel, wrote and asked me to read his book, which I did. After the second review appeared, Webb wrote a letter to *The New Republic* disagreeing with my view of the film's "moral stance" and saying that his greatest objection to the film was that it fails to take such a stance. He based this objection on the fact that, in his novel, Benjamin arrives at the church in time to prevent the wedding, and in the film he arrives after the ceremony. In reply, I said that I had thought at first that Webb's letter was a put-on, possibly by Elaine May; that I didn't understand how the author of this book could equate morality with marriage licenses; and that in structural terms Nichols had tried to improve some of the novel's weaknesses. "Nichols' solution is imperfect, but at least it avoids the destructive cliché of having Benjamin get there Just in Time."

However, there is one point in the novel that I wish had been explicit in the film. Webb makes sure we know that Benjamin is not a virgin when he goes to the hotel room with Mrs. Robinson. I had assumed that Benjamin was not "intact" simply because of his age and kind, but it seems that there were many who did not assume it, and this made a great difference in their view of the first hotel scene. If that is a scene about a novice, it is a conventional skit about initiation; if he is not a novice, then it is about the distress of a young man torn between shock—after all, this woman probably wheeled

THE GRADUATE   45

him in his baby carriage!—and his sexual urges. The conflict between social conventions and (surrogate) Oedipal drives is the source of a deeper comedy.

There were other considerable charges against the picture. Some complained that neither Benjamin nor his parents seem aware that his behavior is not exactly unusual; there is no reference—by California parents—to "Berkeley" behavior or dropouts or hippies. I agree that this is an omission and that it touches the credibility of the environment. Others objected that there was no mention of the Vietnam war; but if there had been "mention" of it, in a film about problems that will persist even if the Vietnam war ever ends, the film would have been accused of tokenism. Still others said that Benjamin was too "straight," that a film about a radical would have been more significant. On this I certainly disagree: what interested me in Benjamin was precisely that he *is* "straight" and that it doesn't protect him, the bottom falls out for him anyway. There would have been less drama, and not necessarily any more social truth, in having these events occur to a member of the SDS. And others have said that the film is not about a real change but about a little rebellious excursion that ends with happy mating and conformity. I don't find this supported in the picture. There is a happy ending, but, as noted, it is a qualified one: none of the things that bother Benjamin is solved by getting the girl, "but, for him, nothing would be worth solving without her." (In the Gelmis interview, Nichols says, ". . . When I saw those rushes [of the ending] I thought: 'That's the end of the picture. They don't know what the hell to do, or to think, or to say to each other.' ") Indeed the film can be seen as testament to the young generation's belief—amply manifested all around us—in the value of romantic love in an arid world.

Anyway, argument about the film's pertinence is quite academic. Box office receipts neither prove nor disprove anything about quality, but they prove something about immediacy; and the financial facts about *The Graduate* are staggering. *Variety* of January 7, 1970, lists "All-time Box-office Champs," rated by distributors' receipts from the United States and Canada. As of that date, the first and second pictures on the list were *The Sound of Music* and *Gone with the Wind,* with $72 and $71 million respectively. Third was *The Graduate,* with $43 million. Third place in only two years—compared with the longer periods that the first two have been in release. Consider, too—which even those who dislike *The Graduate* probably would not deny—the difference in ambition between this film and the only two films in history to attract bigger audiences, and then the impact of Nichols' picture becomes all the more staggering. In the *Saturday Review* article referred to on page 7, Larry Cohen wrote: "If *Blow-up* was instrumental in attracting young people to film, the equivalent American landmark was Mike Nichols' *The Graduate*."

Finally, there was some objection to my remarks about Hitchcock and his tolerance of bad acting. Not everyone will agree, although I have come to do so lately, with Parker Tyler's description of Hitchcock as "a bigtime director of film kitsch," but as for his lack of rigor with actors, think only of Tippi Hedren in *The Birds* and *Marnie* and almost all the principals in *Topaz*.

The Stranger

(*January 13, 1968*)

THE sun—the Algerian sun—was an important part of Albert Camus's early being. It runs all through the first volume of his *Notebooks*. ("The sun on the quays . . . and the port leaping with light.") And it is integral to his first novel, *The Stranger:* the crucial moment of murder occurs when Meursault is in the grip of that same Algerian sun. Luchino Visconti has understood this essential thematic element perfectly. In his color film of *The Stranger,* apparently shot on location, Visconti has aimed to make the sun a benefaction, an oppression, an ambience. One of the world's master cinematographers, Giuseppe Rotunno, who worked on such previous Visconti pictures as *The Leopard,* has helped. The result is a film from which the sun is figuratively never absent; shadows, twilights, nights seem temporary respite from the "leaping light."

This visual realization of the atmosphere is only the beginning of the film's achievements. Visconti has got a faithful screenplay from Suso Cecchi D'Amico, Georges Conchon, and Emmanuel Robles. However, to say that the script is faithful to Camus is both to praise it and to delimit it. It does the most, dramatically and cinematically, that is possible with the book, without any substantive alteration (which would have been intolerable), yet its fidelity gives it the same level dramatic plane as that of films about Jesus. Such films, even one so genuine as Pasolini's, always rise to a plateau and stay on it because there is no hero. A hero must have some illusions and must struggle as a result of them. Jesus is not deceived and will not struggle. So with Meursault.

This gives Visconti's film, like Camus's novel, a quality of observation and patience. The drama is not overt; it is internal, the inevitable abrasion between the protagonist's inner state and the world's protocol. But the film is at a disadvantage in comparison with the novel because of one central matter of technique: Camus evokes a pervasive somnambulistic quality by putting a good deal of Meursault's dialogue into indirect discourse. One example among dozens:

"Why," [the chaplain] asked, "don't you let me come to see you?"
I explained that I didn't believe in God.
"Are you really so sure of that?"
I said I saw no point in troubling my head about the matter; whether I believed or didn't was, to my mind, a question of little importance.

This technique is impossible in film. If Meursault's indirect answers were put on the soundtrack as narration, we would see his lips move as he replied. Or even if his face were not shown, the flip-flop from the chaplain's direct speech to Meursault's narrated replies would have the opposite of the intended effect. It would destroy the texture of the scene, whereas, in the novel, the device creates texture— suspended, dreamlike, *life*like.

But facing the book's difficulties and intent on rendering it authentically, Visconti has made a beautiful, discreet, perceptive film of this epochal work of the twentieth-century Western world. Pictures, in the specific sense, have never been difficult for him; on the contrary, he has tended to indulge himself by slapping pictures all over our eyeballs in films like *The Leopard* and *Sandra*. Here he has *used* his pictorial sense, rather than spewed it. There are plenty of extraordinary things to look at: the skylighted air of the mortuary in an old folks' home, the Algerian streets and rooms (with the smell almost visible), the sensual blending of sea and sun. And when Meursault is in his cell, Visconti (with Rotunno) increases the isolation by increasingly isolating the prisoner's countenance until only his face is embodied out of the dark. But never is the picture merely pretty. Visconti was obviously deeply committed to Camus, and all his previously obtrusive virtuosity is here totally at Camus's service.

Excellently as Visconti has worked, he could not have accomplished what he has done without the art of Marcello Mastroianni, as Meursault. There is a paradox here because, in this French-language film, Mastroianni's lines were dubbed by another actor. The dubbing is well done, and the effect is not as jarring as it might otherwise have been—particularly because most of Meursault's dialogue is flat, meagerly responsive. The performance is made in Mastroianni's face—as he watches from his window on a long Sunday, as he watches his trial from the witness stand, as he watches his death approach him in his cell. Mastroianni was a Visconti discovery; I first saw him as Biff in Visconti's production of *Death of a Salesman* in Rome. Lately he has been walking through some pseudo-Italian Italian comedies. Here, put to the test in a fine role by his mentor (at *his* best), Mastroianni shows us one kind of film acting at its purest: mind and feeling revealed, rather than conveyed, by utmost imagination and simplicity.

Anna Karina, as his girl, responds to Visconti the way she rarely responds to her previous chief employer, Jean-Luc Godard, and gives a live, sympathetic performance. Georges Wilson, Bernard Blier, and Georges Geret (as the pimp friend who starts all the trouble) are exactly right.

It is a truism that the better a novel, the harder it is to make a good film of it. Visconti's version of *The Stranger* is not the exception that proves the rule. Some film adaptations are superior to their originals, which of course this one could not be; some are more or less the equivalent of their originals, which this is not. But this *is* the expression, through their art, by some fine film artists of their sympathy and love for Camus's great book.

Wild 90

(February 3, 1968)

DESPITE the slovenliness and arrogance in his recent books, Norman Mailer has still managed to make clear that there is importance and pertinence in him. No matter how childishly or aggressively or vulgarly he behaved on paper, he has had an innate literary power that he could not shake off. He has no such power as film maker, film editor, or actor. His film *Wild 90,* in which he stars, has only the slovenliness and arrogance.

I cannot say that Mailer was drunk the whole time he was on camera. I do not know it for fact; perhaps it is only colored water that he keeps swilling. I can only hope that he *was* drunk. As for Mailer the editor and director, I hope that he was even drunker. Did he really see those scenes later—possibly when sober—and want to preserve them? It is a frightening thought to anyone who often admired the past Mailer.

Three men—one of them Mailer—pretend to be gangsters hiding out in a warehouse. From time to time they are visited: by the wife of one, by a boxer and his German shepherd, by Mailer's wife. Most of the conversation for ninety minutes (hence the title) is improvised, obscene, fake gangster talk. It is not like Michael McClure's gutter language in *The Beard,* which acted like a wire brush on some of our pretenses. This film's language degrades obscenity.

There will be sages, of film and general culture, to tell us what this picture means. In my view, it is a conversation in a treehouse or a shanty by three boys who are hiding out from their mothers and are cramming all the dirty words they can think of into the ninety minutes before Mother finds them. Along with the incessant "daring" talk, they go bang-bang with empty guns, they punch chairs, they make believe they are tough gangsters, and so on. All of us have done it, but most of us have either got rid of it or have it reasonably under control by about the age of fifteen.

One of the sad aspects of Mailer's power play in our culture (described as such by Norman Podhoretz) is that it has brought him sycophants who would presumably print his laundry lists. One of the sad aspects of the generally happy rise of the film is that it gives facility (in two senses) to this kind of self-abuse of the ego, gives it a spurious importance by making it physically permanent. This is no agonized soul, taking refuge in outlaw fantasy or juvenile dreams as a fortress against existential torment. This is a pampered, possibly drunken little king, taking his bitter little pleasure in making his courtiers hop. The courtiers here are the other players, the cameraman, the sound man, everyone connected with the enterprise who toadies to him. They all seem to me, in this regard, despicable. But Mailer . . . there is pure pathos. In one scene he gets down on his hands and knees and outbarks the German shepherd. That Mailer did it in front of a camera, that he wanted to preserve it, that he wanted to show it in public—well, to some it may be an act of uttermost liberation. To me it is an occasion for mourning.

China Is Near

(February 3, 1968)

MARCO Bellocchio is an Italian director, now twenty-eight, whose second film is his first to get American theatrical release. *China Is Near* is not about East and West; the title is a Communist slogan (the Italian wordplay is lost in English—*La Cina è Vicina*) which, during the film, is painted on Socialist Party headquarters. Nevertheless the picture *is* involved with two worlds—of art. It sets up a tension between symmetry and asymmetry and between the elements of form and of style. Like many of the world's young film makers, Bellocchio has devoured recent cinema developments, particularly French ones; but in this picture the quick-darting camera of impulse, the swift editing that tries to realign logic, the new heel kicking about the fact of film making itself—all these have been applied to a story that might have been contrived by Goldoni. The effect is something like swooping back and forth in a superjet over a formal eighteenth-century garden, piecing the landscape together in free form bits. A good deal of the time it is interesting, but Bellocchio-Goldoni finally has to face down Bellocchio-Godard and beat him.

A wealthy middle-aged brother and sister live in a large house in central Italy. (From a license plate, I take the locale to be Bologna and environs, which also fits the radical politics of the story.) He is a professor who is asked to run for a minor office by the local Social- ists. They assign a campaign manager to him; and this manager is the secret lover of the professor's secretary, with whom the older man is hopelessly infatuated. The professor's well-ripened sister soon gets the manager into her own bed, and when the secretary discovers this, she gives herself, in revenge, to the importuning professor. The comedy is always oblique and dry—never dryer than the scene, early one morning, when the manager leaves the sister's bed as the secretary leaves the professor's bed, and these two—themselves lovers—finish dressing in the hall and steal away together out of the house. Both of the women get pregnant and eventually resign them- selves to having their babies. The two couples will apparently live together; what their sex life will be in that house is anyone's humid conjecture. But could any story, or conclusion, be morally neater, more thoroughly based in cheery eighteenth-century rationalism?

There is also a fifth prominent character—the younger brother of the wealthy pair, a supersober communist who moves around the quadrilateral gavotte like a mundane Savonarola. The best character touch in the film is that this boy himself is satirized, rather than being used as a heavy truth-symbol posed against falsities.

Presumably because of his youth, Bellocchio seems to have felt overobligated to employ up-to-the-moment methods. They mislead us, they make us expect a different kind of film. When a nice old plotty comedy begins to be visible, we feel a little irritation with the somewhat irrelevant Nouvelle Vaguery. An older director—Monicelli, Comencini, or the earlier De Sica—would have gone straight for the story and wrung the last drop of juice out of it. Bellocchio's approach produces some longueurs. At its most germane, it makes comments that would be more difficult in traditional style: telling us that the truth and foolishness of human feelings are perennial; that politics —even radical politics—shares both the truth and the foolishness, perennially. At its most disjunctive, the style seems just cinematically hip.

I saw Bellocchio's first picture, *Fist in His Pocket,* at the New York Film Festival a few years ago, but will withhold comment until its public release, which is due soon. But I cannot withhold comment on the aptness of the director's name. Even an Italian Dickens would not dare to call a film director Mr. Beautiful Eye.

Love Affair

(February 17, 1968)

INTERESTING times in the film world. As I've noted here, the extensions in film language, in film imagination, that appeared in France during the last ten years have been gobbled up hungrily by many young film makers around the world. There are historical reasons why these extensions were inevitable (and they are not all happy; one of them is a kind of intellectual and artistic sloth); but a new vocabulary can hardly be good or bad in itself, everything depends on its use. The chief shadow in this new school has been the speed with which the new liberation has in some cases become imitation, even parody. This is the risk that every innovator faces. Imagine Hemingway reading Alfred Hayes. Imagine Godard seeing Loach's *Poor Cow*. Whatever the faults of the originator, he is not responsible for his mimics.

But some young directors have done better than imitate, even skillfully; they have absorbed the new language as part of their mother tongues, as a chance to be more fully themselves. I have seen none of whom this is more true than a young Yugoslav named Dusan Makavejev, who has written and directed *Love Affair, or The Case of the Missing Switchboard Operator*. This is his second feature film; the first has not yet been publicly released here. In *Love Affair* the New Wave influences have been thoroughly assimilated and are integral to the director's material, something I did not feel in *China Is Near* by the admittedly gifted Bellocchio.

Style is content here. What is the story? Isabela, a switchboard operator in Belgrade, meets a sanitation inspector named Ahmed. They have an affair. She loves him, but she is sexually ductile. While he is away for a month, she is seduced by a postman and becomes pregnant. As an act of atonement, she lets Ahmed think it is his child but that she doesn't want it. He is revolted and leaves. She pursues him, finds him drunk (he is not a habitual drinker), and pleads with him. In shaking her off, he accidentally kills her— knocks her into a deep underground well. He gets drunker, collapses, and is soon found by the police.

What a trite and tritely sordid story. What a charming, light, poignant, and socially illuminating film. Makavejev has sketched in a good deal about life in Belgrade today, has invested his little film with a highly personal view of fate and fate's ludicrousness, and has directed so well that the texture itself gives us a sensual pleasure.

The picture opens with an amiable old sexologist lecturing us, as he might do on educational television, about sexual customs and attitudes (except that some of the drawings we see would burn out our TV tubes). The old man reappears occasionally through the film, never saying anything directly relevant to it. He supplies a note of frankness about sex which is quaintly old-fashioned against the realities for which mere frankness is insufficient.

The chronicle of the lovers, as such, is handled with fine astringency. For example, we never even see them meet. She is out walking with a girl friend, then we see these two girls crossing the street with a man whom—presumably—they have chatted with while waiting for a traffic light. One appeal of this picture is the "presumably": it is easy and pleasant to fill in the gaps of omitted detail. The touches of characterization—and their social relevance—are supplied with similar casualness. Ahmed, to judge by his name, is one of Yugoslavia's numerous Mohammedans, therefore a man who takes fidelity very seriously; he is, moreover, a very serious communist and therefore additionally puritanical. Isabela is a foreigner, a Hungarian, and always conscious of it: she sings Hungarian songs, bakes a strudel, talks about her "Hungarian" need for sex. The fact that Makavejev made his heroine a Hungarian may be his comment on the Yugoslav attitude toward Hungarian views of the seriousness of communist life.

Structurally the director has folded the story back on itself so as to thicken its texture. No sooner have we met the pair than we flash *forward*—to the police fishing the girl's body out of the well into which she falls at the end; and all through the film there are intercut sequences of her autopsy, as well as a criminologist's lecture on the nature of murder. Before we see the girl naked for love, we have seen her naked on the autopsy table. Before we see her black cat stroll lazily over her rounded white bottom (a lovely picture), we have seen the autopsy instruments lined up on her naked legs. The first intercut of the death sequences is puzzling; then we understand that a linear event has been sliced and the sections intertwined so that we could watch all of it, figuratively, at the same time. The effect is not only of the multiplaned simultaneity that has been a part of modern art since Picasso, but a reminder of the fragility of life, and, more, of the fragility of life patterns. And implicit in this is a comment on the final irrelevance of political systems to some matters of biology and the psyche.

Makavejev's use of cinematic resource is effervescent but careful. There is a shot of the pair at a resort hotel—he on a little balcony, she opening a window next to him—that is as exquisite as the window-opening scene in the morning sun in *Jules and Jim*. While Isabela is rolling the strudel dough, an operation rich with five hundred years of Middle European social history, the sound track gives us the Triumphal March from *Aida*. When the postman is working her over at the switchboard and she begins to feel her glands moisten, we suddenly cut to some films of nude *tableaux vivants,* c. 1915, in which a moustached gentleman and an ample lady pose on a revolving turntable as Adam and Eve, etc.—a touch that both conveys and mocks lubricity. When Ahmed gets drunk, Makavejev follows him with a hand-held camera; it is the only time he uses this device and thus there is some sense in it, for once. The very last shot in the film, after Ahmed's arrest, is of the front of the resort hotel where once they had been happy, as we hear an East German Party song on the sound track—a record that had been sent Ahmed by some friends and that he had played there on their "honeymoon." The contrast between his rigidity and her fluidity, of which this last moment reminds us, in a strange way make us feel more confident. If there are elements in human behavior that can never be controlled or predicted, then we can acknowledge them and not be depressed by them. Ahmed will still be murdering Isabela in any society we can dream of; so we can just take that for granted and get on about our business of revision or revolution or research or relaxation, whatever our bent happens to be.

A Hungarian actress named Eva Ras, plain of face and enchanting of smile, is completely winning as Isabela. It is quite pertinent to this film of contradictions that her body is surprisingly beautiful when she undresses. Ahmed is played with fit stolidity by Slobodan Aligrudic, a name I have not invented. It is also relevant that this film runs only seventy minutes. Many "new" films seem interminable—endless wandering through streets, pointless conversations, pauses meant to be pregnant that are usually virgin. Makavejev has made every second of his seventy minutes count. The result is optimal: his film is never tedious and yet it is long enough to be satisfying.

Tell Me Lies

(March 2, 1968)

PETER Brook's film against the Vietnam war is the wrong film at the right time. The filth—of outrageous government falsehood, of responsibility for pointless death and destruction—is piling up in America in a way that makes this country seem a very great deal worse than New York during the garbagemen's strike. If we are to have a big Eastmancolor picture on this subject now, it ought—in heaven's sweet name—to be one that heightens and thrusts forward all possible American opposition to the war. Which means, for the most pragmatic reasons, not for dainty esthetic ones, that it ought to be a good film. *Tell Me Lies* is not.

A couple of years ago Brook produced an antiwar theatrical work called *US* with the Royal Shakespeare Company of London, parts of which were included in a short film called *The Benefit of the Doubt,* directed by Peter Whitehead, that was shown at the last New York Film Festival. Now Brook has used *US* as the starting point for a long cinema fantasia on the same theme and has used most of the same actors. (Note, to our additional shame, that both of these films were made in Britain. The only major American picture I know of on this subject is John Wayne's forthcoming *Green Berets,* whose aim will presumably be different.) *Tell Me Lies* begins with Mark, a young Londoner, becoming upset by the magazine photo of a bandage-swathed Vietnamese baby. He sets out on a kind of pilgrimage around London, asking questions, attending rallies, having conversations with known and unknown figures. There are also reenactments of some actual events, like the self-immolation of Norman Morrison, the American Quaker, before the Pentagon. There is a portion of the (New York) Open Theater's ribald playlet on how to beat the draft. There are songs and sketches. There is a general air of tension.

But it is an air of theatrical tension, of self-dramatization. After Mark and his girl leave a cinema where they have seen a Buddhist monk burning himself in Saigon, she asks him, "What is there that we would be prepared to burn ourselves to death for?" The question seems to me typical of what is wrong with this film—thin dramatics over a void of thought. The answer to her question, I should think, is: "Nothing, I hope. And I really don't feel remiss about it, either. But I also hope there are some things I would want to stay alive as long as possible for—and to fight about. There are far too many burners in the world already. Why do their work *for* them?"

The attitudinizing songs, the soul-baring professions, the "daring" sketches, are all in a mélange of Brecht-Artaud modes that is supposed to assault our minds and our nerve ends, but it only distracts us to the *surface,* to the actors themselves being personal and rather designedly uninhibited; so we lose sight of Vietnam in a welter of uninteresting candor, as well as of theatrical theory. I am not much interested in sophomoric political discussions in Chelsea basements simply because the familiar juvenilities are couched this time in British locutions. I am not the least bit interested in the actress Glenda Jackson being vibrant at me with banal discoveries she has made in the recesses of her soul. (I would be *very* interested in seeing this fine actress—who was Charlotte Corday in *Marat/Sade*—in a good script that dramatized the putrescence of this war.)

Earlier, at an outdoor rally, Miss Jackson reads a ringing statement, and when Mark asks her who wrote it, she says quietly, "Ché Guevara," and passes on. We are meant to reel, some way or other, without any questions at all, silenced by that charismatic name. This is totemism, not politics.

And this leads to a terrible irony. I think that Brook's film works just against the crusading effect he meant it to have—precisely because of his cavalier treatment, or dismissal, of politics. This was crystallized for me in one incident. At a party Mark converses with Kingsley Amis and Peregrine Worsthorne, two of Britain's staunchest supporters of Johnson's Vietnam policy, who are clever polemicists. Their arguments about stopping communism and drawing the line in Vietnam were for me the points that were made most strongly in the whole film! I emphasize as loudly as I can: they are points with which I disagree. But the rest of the film is so woozy or quivering-souled or *East Village Other* bawdy about matters I agree with, that the Amis-

Worsthorne views stand out as a (deceptive) moment of clarity. I
would have given all the songs, all the blue jokes, all the squiggles of
soul searching, for one minute of cogent reply to their views. The
Amis-Worsthornes need to be *answered,* especially to convince the
unconvinced; and pictures of charred babies will not move their supporters. No one displayed pictures of the charred babies of Hamburg
in 1943 or of grievously wounded Egyptians in June 1967—at least
not as arguments against the Allies or Israel. Why? Because the horrors of war were taken as necessary for political ends. Well, that is
exactly what the Amis-Worsthornes of the world are saying now,
and all the pictures of blasted villages and burning Buddhists are only
clucked over as part of the price for necessary ends.

Two other recent films about Vietnam also fell far short of the point
for which they were presumably made. *The Anderson Platoon,* by the
Frenchman Pierre Schoendorffer, was a documentary about some U.S.
soldiers, their courage and their hardships. Some of the footage was
fine, but it said nothing about *this* war; with a change of helmets and
rifles, it could have been made on Guadalcanal. Felix Greene's *Inside
North Vietnam* showed us in color how lovely the Vietnamese
countryside is and how graceful and winning the people are, even if
the cheerful workers were possibly under the eye of an off-camera
commissar. But all it accomplished that was really relevant was to re-
fute the American contention that we have never bombed North Vietnamese civilian centers. And Greene completely avoided the political
issue that is the *reason* for the war; his sound track never once mentioned the word "communist."

It can be argued that any film that pushes the daily ghastliness of
this war into people's faces is worthwhile. But many of those who
support the war know about the ghastliness quite as well as the war's
opponents. I cannot imagine that *Tell Me Lies* would change a war
supporter's mind; and unless more and more minds are changed, the
war will not quickly end.

In my view, what we need most are not products like *Tell Me Lies*
or *US* or *Viet Rock,* which are, finally, coterie works; and we certainly
do not need blitheness about some hard political issues as in *Far from
Vietnam,* a French film that was also shown at the last New York Film
Festival, some of which assumed that the problem of communism
only bothers squares. We need some first-class film documentaries on
the political answers to the Amis-Worsthornes: documentaries that,

among other things, tell the truth about the belated veneer of purpose on an initial misadventure; on the escalation of military mistakes and political falsehood; on the impossibility of achieving the very aims our government professes (because we either have to occupy Vietnam permanently or else submit *after* a treaty, rather than before, to its Communist domination); that the "line" has been drawn in the wrong place. Such documentaries would change those of the opposition who *could* change much sooner than pictures of charred babies. If enough people changed, then politicians—most of whom are essentially amoral—would respond; and it is only through the response of those with power in Congress and elsewhere that matters can improve. Take an example from the lower end of the integrity scale in politics— Richard Nixon. If a sufficient number of people expressed opposition to the Vietnam war tomorrow, it is a safe bet that Nixon would make his next campaign speech under a huge photograph of Dr. Spock.

Accattone

(April 6, 1968)

PIER Paolo Pasolini's first film, *Accattone* (1961), gets its American theatrical premiere long after it has, quite literally, found a place in cinema history. (Several books of the last few years discuss it.) His later films, *The Gospel According to St. Matthew* and *The Hawks and the Sparrows,* have already been shown here. I saw *Accattone* in 1964 in Rome, again at the 1966 New York Festival, and again recently; and, for me, it lives as a work of narrow but intense vision—a film about viciousness and criminality that evokes compassion. Its style is neorealist: it was made on locations, not in studios, with nonprofessional performers. Sometimes this method makes merely vernacular films, but it gives *Accattone* a grainy, gripping authenticity.

Pasolini came to films after winning distinction as a novelist and poet. He won the Viareggio Prize for poetry in 1957 with *The Ashes of Gramsci.* In the following year Sergio Pacifici called Pasolini "one of the youngest and most mature poets to come to prominence [in Italy] after the last war," then went on to say:

Certainly the reference to Gramsci, the founder of the Communist Party in Italy, is more than a mere tribute. . . . It was Gramsci who, in the late twenties, writing from the jail where he was to die, urged the formation of a new culture that to become "popular" must reflect the aspirations of the people. . . . Pasolini has already done much to narrow the gap that has always existed between life and literature in Italy.

I had the chance to speak with Pasolini on television in New York in 1966 and took up this point. His use of dialect in prose and verse was obviously an attempt to make literature "popular" in the Marxian sense; was this the same impulse that had taken him to neorealism in films? Yes, he replied, but more than that, the results had led him to abandon literature for film making, at least for a time. In his writing (I paraphrase from memory) he had sought the quintessences of factualness. Film gave him the power of fact to *start* with, and he could go on from there. In his two best films so far, *Accattone* and *St. Matthew,* I think this is precisely what he has accomplished.

Accattone, the hero's nickname, means "beggar." (The dialogue is in Roman slum dialect, which, I am told, many upper-class Romans have trouble in following. This effect—untranslatable—makes the slum a segregated province, moated and clannish.) Accattone is a pimp and spends most of his time lounging about with other young men who apparently are also pimps. Their scenes are like big-city criminal versions of the loungings of the aimless small-town youths in Fellini's *I Vitelloni.* Accattone's girl, whose prostitution supports them both, is jailed—a considerable term, for perjury. He has no money, but it is a principle with him, as with his companions, not to work for a living, the way his brother does. Accattone nearly starves. His friends respect his fidelity to principle at the same time that they do nothing to help him; in fact, they taunt him. (This, too, seems part of the code.) Then he finds another girl, seduces her, and induces her to try whoring. Because she loves him, she attempts it but cannot go through with it. He is now so emotionally involved with this girl that he sacrifices his principles and tries a job, but *he* cannot go through with *that.* He turns to thievery as a means of keeping his girl off the streets, is chased by police, jumps on a motorcycle, and is killed in a collision.

This synopsis may suggest a tract about society forcing criminality on the poor, of pimps and whores as pawns of capitalism's ruthlessness. Pasolini is too good for that. Certainly the film is aware that bourgeois society needs prostitutes as reverse endorsements of its virtue, just as it needs thieves to endorse the sanctity of property; but Pasolini is a Marxian *artist*. His Marxism directs his sympathies, then his art takes over. His people exercise options as completely—if not as widely—as anyone else. Accattone lives as stringently by his code as any parfit gentil knight. He embraces his small son to steal the boy's medallion and pawn it to dress his girl for her trade, but he does it with an air that says the child would understand if he were old enough. No facile tears for the victimized poor. Accattone is not much more a victim than most of us, and he has more pride (though inverted) than many non-pimps.

As for Pasolini's direction, its most remarkable feature is that, although this film is now seven years old—seven years of accelerating stylistic innovations—it is neither up to date nor old fashioned. It is a piece of straightforward, traditional, intelligent film making. There are no "poetic" shots—nothing remotely as stunning as the elevation of the cross in *St. Matthew*. Pasolini's strength in *Accattone* is not in fancy camerawork or in editing but in his almost violent closeness to his material. He selects and states—simply, fiercely. The simplicity conveys the fierceness. Yet the picture is not spare: it sits in a nice full texture of these people's rites and habits.

Pasolini has chosen his cast excellently—for individual flavor and balanced colors. Franco Citti, the Accattone, has a blunt, unforgettable face, square-jawed yet with the requisite weakness, a man whose self-pity flows so readily that it makes us pity the man who needs self-pity so badly. Franca Pasut, his (second) girl, is heavy, servile, pretty, very moving in her devotion to Accattone and her remorse at not being able to lay strangers in order to support him. The minor characters are chosen like gems by a jeweler: Mario Cipriani as a turkey-cock thief; Umberto Bevilacqua as a Neapolitan hood whose beetle-browed, broad smile is scary; and an anonymous, runty, wide-eyed girl as the bereft wife of a man in jail, with a brood of kids who move around her wherever she walks like an animate hoop skirt. Most of the characters, like most lower-class Italians, burst into snatches of irrelevant song as they walk or idle, even as they scheme.

Pasolini handles the pimp's discovery of love without sentimentality. There is one risky point, but he redeems it. At a café on the Tiber, when a stranger sends a waiter to pick up his girl (having assumed that she was available), Accattone assents, largely because his friends are watching. As he sees the stranger fondling her, he suddenly announces to his friends that he's going to leap from the bridge —something we have seen him do earlier to win a bet. But this time he is drunk, and his friends run after him to restrain him (*laughing as they do so*—a masterly touch). They pull him down, he runs to the water's edge, wets his face, then rubs it in the sand. For a split second, the self-debasement seems too obvious. But the closeup of Accattone's sand-plastered face is so ugly—the ugliest such shot that I know since Charles Vanel's face went into the mud in Clouzot's *The Wages of Fear*—that the moment is purified.

The music on the sound track is as unsatisfactory as it is in *St. Matthew*. Pasolini bastes on Bach at deliberately inappropriate moments—as during a fight between Accattone and his estranged wife's brother. The purpose, I suppose, is to assure us that in these struggling animals are souls as precious as any pictured in that music, but it seems affected. And the very end of the film seems strained. We see Accattone with his head against the curb where he has been thrown. He murmurs, "I'm all right now," and dies. It is hard to believe his acceptance of death. A man who has lately had a bad dream of his own funeral? A defensive dramatizer of his right to exist? A man who has, for the first time, found a girl he does not want to exploit? His resignation seems Pasolini's, not his own.

But *Accattone* sticks in the mind—small, stubborn, vivid. It is credible, not pat; hard, not tough; humane, not lathered with soapy social significance. It *uses* its facts, acknowledging that the film form itself can make them real and that therefore the film maker has an obligation to take us inside factualness, where we can see the muscles coiling. And this, as Pasolini said, is why he makes films.

POSTSCRIPT. If only that was why he had continued to make films. But his subsequent works, including *The Hawks and the Sparrows* and *Teorema,* have become increasingly allegorical, increasingly picturesque in the worst self-conscious sense, increasingly grandiose in a soft, emasculated way. When Pasolini was making films about and with the people who used to be his subject in fiction—gutter folk—his work had vitality and contact. As his films have gone up the scale socially, they have become inflated and aloof. The only element in *Teorema* that had any interest for me was the story of the serving girl.

Hour of the Wolf

(April 20, 1968)

INGMAR Bergman's previous film, *Persona*, was related to his earlier work thematically and superficially. It dealt with two women, one of whom had a young son, which gave it a similar "orchestration" to *The Silence*. Its seaside setting and its quality of light connected it with *Through a Glass Darkly*. The character of the actress had the same name (Vogler) as the actor in *The Magician*, and the very title *Persona* (mask) complemented the original title of that film (*The Face*).

In *Hour of the Wolf* Bergman continues his systems of linkage. Like *Persona*, the new picture is about two people living by the sea, one of whom is mentally ill. Liv Ullmann, who played the sick person in *Persona*, is the healthy person here, and she has the same name (Alma) as the nurse in the last film. Her lover, the sick person, has the same surname (Borg) as the old doctor in *Wild Strawberries*. The name Vogler is again used, for Borg's ex-mistress. There is even an attempt—an afterthought, it seems to me—to carry on the "film consciousness" of *Persona:* under the credits we hear the chatter on a studio set, then the warning buzzer sounds, the chatter dies, and the film begins. (This element—of film consciousness—is not used further.) Some of these references to past work seem only a private game; some of them seem intended to bind *Hour of the Wolf* in two ways to all the serious films Bergman has made since *Through a Glass Darkly*. In form and tone, it is a "chamber" film—a term Bergman obviously derives from Strindberg's *Chamber Plays,* which, as Jörn Donner says, "cover short spans of time with few actors and possess something of the character of intimate music." In theme all these chamber films are concerned with mental anguish.

Most of *Hour of the Wolf* is a flashback. It begins seven months after the death of Johan Borg, a painter in early middle age. His pregnant young widow, Alma, comes out of their island house, sits, and talks to us. (She knows Johan's internal experiences through a diary he has left.) Then we see Johan and Alma arrive at this island off the Swedish coast seven months before. He is in a poor mental state, worsens, goes quite mad, shoots Alma, thinks he has killed her, then goes off into the woods and kills himself. Other characters figure in the story, in fact and in fantasy. An impoverished baron lives with his wife and relatives in a castle on the other side of the island. The baron is an admirer of Johan's and has one of his paintings—a portrait of Johan's ex-mistress, Veronica Vogler. Johan and Alma are asked to dinner, and in the course of the evening there is a scene—reminiscent of that between the novelist and his rich host in *La Notte*—in which the painter tries to justify for himself his life in art. He says he is in the grip of a compulsion—art—that is irrelevant to most people in the world. (Something Bergman has said of himself at various times.) Later, there is a long sequence of hallucination in which Johan fancies that he revisits the castle and meets Veronica, whom the baron has invited so that Johan can copulate with her while others watch. This hallucination presumably derives from the fact that the baron owns the portrait of Johan's ex-mistress, the memorial of the artist's passion. It is a deranged comment on an artist's inescapable exposure of privacies.

Some attempt is made to break out of the confines of case history —principally with Alma, who takes Johan's condition as her spiritual responsibility. Near the close she even has a hallucination of her own, as if in an effort to join him. (Another resemblance to *Persona*.) And at the very end she tells us that there must have been something more she could have done to save him. But this strikes us as her compassion, not a truth of the case. The film records the progress of a sickness, and the most we can feel is a hospital visitor's pity. Much of the time the picture is somewhat clinical and remote because Johan does not represent us.

We are given no reason to believe that any of the things in the world that might drive *us* mad are relevant to Johan's condition. *Persona,* which I think is Bergman's best film and his masterpiece, is about a woman hounded by the horrors—curable and otherwise —that hound many of us. She flees by inner withdrawal (a compromised suicide); and then the film contrasts her withdrawal, which seems starkly rational, with the nurse's irrational health. But we know little of Johan's past except his affair with Veronica and the fact that he has a son. There is nothing to persuade us that we are watching anything but the course of a disease. In *Through a Glass Darkly* the girl's madness is bound closely to the theme of spiritual hunger, both in her own agony and in the colors of the characters around her.

But Johan, for all we know to the contrary, would have the same disease and would suffer the same way in a religiously secure world.

Willy-nilly, then, Bergman assumes the responsibility of making the *images* of Johan's illness so graphic and moving that they hold us. For the most part they are surprisingly weak—given Bergman's talent and theme—and almost predictable, except that in some of the hallucinations he seems to have used a different film stock for heavier black-and-white contrasts. (As in the scene where Johan murders a boy who is presumably his son.) The distortions, the nightmare choruses, the wild projections—all these are just about what we would expect. Even the work of Sven Nykvist, who has been the cinematographer on all but one of Bergman's films since *The Virgin Spring,* is not as beautifully and softly superreal as it usually is.

But no Bergman picture is barren. He begins here with Alma facing us and talking to us—for at least three minutes. It certifies his oft-stated belief in the human face as the theater of life; and the very fact of his self-confidence, that he doesn't feel obliged to hop about to keep the shot from being static, *keeps* it from being static, makes it quietly daring. Later there is a scene in which Johan times one minute of silence by his watch. Bergman makes a drama of those sixty seconds by conveying Johan's triumph in getting through at least one more minute of his life sentence. There are some pure virtuoso touches, like one in which Johan is seated on a stony beach, painting. On top of a rise a girl appears, with only her legs visible. Her head is kept outside the frame even as she approaches. We assume it is Alma. Suddenly, when she is next to him, she kneels and we see that it is another girl. (Veronica, as we learn.) It is more than a surprise, it is a revelation. Because Veronica never really comes to the island, the surprise tells us retroactively that we have been watching an illusion.

Further, there are extraordinary performances in the two leading roles. Max von Sydow, the Johan, is one of the world's best screen actors. One proof: he is not only good in good pictures or mediocre ones, he is good in bad ones; witness *The Greatest Story Ever Told* and *Hawaii.* His power to concentrate every atom of understanding and imagination and presence on every instant is what keeps Johan from becoming dreary, although von Sydow cannot by himself make the character relevant to us. Liv Ullmann, the virtually mute actress in *Persona,* will win any viewer but will doubly impress those who saw the previous picture. With no flash trickery of makeup or accent, with no equipment but empathy and a talent for truth, she creates here an utterly different character. The aloof, enigmatic, elegant actress of *Persona* is transformed into the fresh, vulnerable but wise girl. This is creative acting, as distinct from vaudeville impersonation.

But, unfortunately, there is a good deal to dislike. The scenes in the castle, including the long final fantasy, alienate us, when they should involve us. All the anticipated or unfruitful symbols, the close-ups of faces that seem labeled D for Decadent, the unsubtle chiaroscuro, are a disappointing rehash of old German expressionism, something like Mai Zetterling's *Night Games* and *Loving Couples*. These castle scenes, as well as other failed introspective scenes, make us conscious of cinematic strain, rather than taking us on a painfully intense journey through a mental hell.

The hour of the wolf, says Bergman, is the hour between night and dawn "when most people die . . . when nightmares are most palpable." It is the hour when Johan cracks irretrievably. (Another sign of Bergman's strain: he inserts the title again just before this last sequence to make sure we get the point.) The shadow world of mental-spiritual torment is Bergman's own, marked out by him more and more devotedly as he goes on; and the wolf hour is high noon in that world, therefore of inevitable interest to him. But his specific material here is much less resonant than in his past chamber films, much more clinical, and I think that the evident struggle of his film to *be* a film reflects—subconsciously, perhaps—his sense of this aridity.

Well, in the long run, what of it? Bergman's films have fallen short, to some degree, at least as often as they have succeeded, principally because he ventures more deeply into the shadows than most directors. *Hour of the Wolf,* his twenty-eighth film, is not one of his successes. That is all. Bergman lives and works—happily, is "compelled" to work. I look forward to his next.

La Chinoise

(*April 27, 1968*)

JEAN-LUC Godard's new film is called *La Chinoise,* but the Chinese
girl of the title is French: a Parisian student named Veronique who
is devoted to Mao and Red Guardism and thus is not only anti-
capitalist but anti the French Communist Party and the Soviet Union.
She shares an apartment with her lover and another young couple
and an extra man, all of whom also share her views. The extra man
is named Kirillov, but *The Possessed* has been used as a launching
pad, not as a model. To Godard's credit, he understands that Dostoevsky sees the buffoonery in his serious characters. This buffoonery
suits Godard's familiar methods; and *his* characters' ideas, highly
debatable though they are, at least give *La Chinoise* a unity that is
rare for him.

Godard plays. That is the only way he knows how to be serious.
(In principle, a wonderful gift.) He plays with almost every con-
ceivable cinematic device. He plays with the world's literature. (There
is the usual welter of references to authors, and Veronique's lover is
named Guillaume Meister—an actor, of course.) He plays with art.
(Picasso is invoked with a bull's head made of a bicycle seat and
handlebars.) He plays with suicide. (Kirillov shoots himself as if his
death were only an incidental footnote in his life.) He plays with
murder. (Veronique kills the wrong Soviet embassy official, then,
as if she had forgotten her gloves, goes back and kills the right one.)

And he plays with politics. His characters use it as sport, as sexual
arena, as décor (literally: the walls are painted with Maoist slogans,
the shelves are pretty with dozens of little red books), as a verbal
football in consciously useless debate. They walk around reading
Mao aloud like monks with Scripture. Al Carmines, the minister-
composer at the Judson Church in New York, recently wrote a one-
act "opera" on Mao's sayings that used this phenomenon better.
Carmines' musical settings understood what the maxims say but also
understood the pleasure—the diversion—of incantation. Godard's
characters seem to think they are engaged in political action when
they walk around the apartment reading; and although he thinks

there is some amusement in his revolutionaries, he is certainly not satirizing them.

There are only patches of story in the film, and those patches are left shredded. (What happened to Veronique after her killings?) Structurally, it cannot be discussed in conventional terms, only in terms of *jeux d'esprit*. Often the *jeux* do have *esprit:* of youthful spontaneity and non sequitur, of "inside" cinema references along with nose-thumbing at the canons of cinema, of a "straight" young-love scene (Veronique and Guillaume and a record player) or of an equally conventional "student" scene. (One couple is sleeping in bed, the other couple on chairs, while the radio blares the *Internationale;* Kirillov walks across the bed, picks up the radio, walks back across the bed and out, while the two couples do not stir.) And throughout there is striking color. Raoul Coutard, the photographer of all but one of Godard's pictures, has framed these people against great plaques of red and blue and white (flag colors) and has transformed shot after shot into posters.

Some of the performers are taking. Jean-Pierre Léaud, who was the child in Truffaut's *The 400 Blows* (1959) and graduated to Godard's *Masculine Feminine* (1966), plays Guillaume with a nice fiery scruffiness. Michel Semeniako plays the other male lover, who is "expelled" for revisionism, with a homely, sad self-reliance. Godard, as he has done before, also uses an actual personage: in this case Francis Jeanson, the noted radical writer and teacher, who is very agreeable and who has a long conversation with Veronique—nicely shot on a suburban train that stops every twenty seconds or so. The conversation has a feeling of mutual respect and mutual pathos, because he cannot convince her that isolated terroristic acts are futile and she cannot convince him that, for her, futility is not necessarily a deterrent. The one weak member of the cast is Veronique herself, played by the latest Mrs. Godard, Anne Wiazemsky, whose acting ability is not apparent and who resembles a depressed goldfish.

But if *La Chinoise* is one of Godard's more organic films, it is still faulty and reprehensible, in terms of what it does and—eventually— what it says. (And if we are told that the two matters are inseparable, this really does reduce Godard's politics to a game—merely an occasion for style.)

Take it first purely as film. Despite all the high-spirited high jinks and the glints of charm, the net effect is somewhat cold. In *The Married Woman* Godard chose the subject of sex and rendered it as a lovely, absolutely unerotic, marmoreal ballet. Here he treats the fuming of young activists in such a disjointed and fragmented way that we observe, rather than sympathize; we watch for the next bit of horseplay, as at a madcap vaudeville. His own cleverness, up to the split second though it may be, intervenes. The beetle-on-a-pin feeling is heightened, rather than otherwise, by those fake *cinéma vérité* interviews that Godard likes. He puts his actors against a wall, one at a time, then asks them questions (his voice barely audible), which they answer in character. The first few times he used this device in his pictures, it seemed to promise greater immediacy, a broadening of his character almost in an essaylike way. Now the novelty has worn off, and we are conscious only of his intrusion into the *life* of the characters, his insistence on branding them and his work with a trademark.

Out of all the rest of the incessant barrage of devices, look at one more. Godard frequently reminds us that this is a film. Before some shots we see the clapperboard marking the take; sometimes the actors address the camera; sometimes we even see the camera, with Coutard behind it. When Richard Lester's actors in *How I Won the War* acknowledged that they were in a film, Lester was telling us something about entrapment, about roles forced on us by the juggernaut of social forms and human stupidity. With Godard it is only a gifted man's high-spirited impatience with the limits of his gifts and of his medium.

One incidental note. All through the film he cuts in swift shots of American comic book heroes as icons of contemporary violence. The rabid pro-Americanism of young French artists and intellectuals in the fifties has now turned to rabid anti-Americanism. But the subjects at issue are *exactly the same:* comic books, gangsterism, toughness. Vietnam (and possibly Kennedy's murder) turned the abstraction of American violence into a reality. When it was only a game, French youth liked it; but not now. The lesson seems to have been lost on this film's young political activists.

Which leads to the second principal matter—what the film says. In the dialogue with Jeanson, Veronique tells him that she wants to blow up the university or the Louvre as an act of protest, and he tries to dissuade her. She reminds him of his opposition to the Algerian war and how he suffered for it; he reminds *her* that this was not an isolated act of self-gratification. He makes no impression on her. For her, politics remains personal thrill, romance, vengeance.

I wish I could believe that Godard was commenting on her views and those of her friends, that he was saying by implication: "Reading Mao bravely in a Paris apartment—or an American dormitory—is great defiant fun, but it is cheap fun. To take it as more than fun either shows a failure of imagination or a dogged romanticism that wastes potentially useful radical energy. And if you tell me—even tell me seriously—that to live under Maoism would be no worse than the way things are now, my reply would be, Why bother? All that turmoil just to establish a different Establishment?"

Or, failing that, I wish I could believe that Godard is a committed Maoist; it would at least give the picture an internal validity. But the impression grows and persists that Godard is congenitally a bootlicker of young boots. When he made *Breathless* almost ten years years ago, postwar nihilism was "in" with youth, so it was "in" with him; *Breathless* was nihilistic. (It was his first film, and it seemed a personal statement.) Today the cognate youth group is politically activist, so *La Chinoise* is Maoist.

I do not suggest that Godard is forbidden to change or that he should distort the current state of society or should abandon his interest in youth—which is really his only interest. But from the course of his work one may deduce a consistent belief: Young equals Good, Older equals Bad. This simplistic tenet—held by some of the Older, too—is particularly disquieting in *La Chinoise* because of its subject. Many of us agree that the world is desperate for social change, for radical shifts in values, but not all of us think that the Maoist "solution" is glamorous. In any event, it is not a subject that can be fulfilled in play or in charm—even in Godard's somewhat frigid charm. Either *La Chinoise* is fundamentally serious or it is an inconsequential divertissement on a serious theme. If the latter, it is irresponsible; if the former, it is glib.

At one point Guillaume acts out the story of a Chinese Maoist (in Moscow, I think it was) who appeared before the press with his face wrapped in a towel and complained of a beating. He slowly unwrapped the towel (as Guillaume does) while he talked, revealing at last an unharmed face. The reporters were confused; but Guillaume explains to us that it was wonderful political theater. It *is* momentarily effective; but very soon the juvenility of the Maoist action and of Guillaume's endorsement (via Godard) seems trifling. The scene epitomizes the film.

POSTSCRIPT. Some have said that the subsequent student revolt in Paris in 1968 proved the worth of *La Chinoise*. I can't see the connection. There will probably be space stations in 2001, but that fact won't improve Kubrick's picture as such. (See following.) In neither case did I question prophetic powers.

2001: A Space Odyssey

(May 4, 1968)

STANLEY Kubrick's *2001: A Space Odyssey* took five years and $10 million to make, and it's easy to see where the time and the money have gone. It's less easy to understand how, for five years, Kubrick managed to concentrate on his ingenuity and ignore his talent. In the first thirty seconds, this film gets off on the wrong foot and, although there are plenty of clever effects and some amusing spots, it never recovers. Because this is a major effort by an important director, it is a major disappointment.

Part of the trouble is sheer distension. A short story by Arthur C. Clarke, "The Sentinel," has been amplified and padded to make it bear the weight of this three-hour film (including intermission). It cannot. "The Sentinel" tells of a group of astronauts who reach the moon and discover a slab, clearly an artifact, that emits radio waves when they approach it. They assume it is a kind of DEW marker, set up by beings from a farther planet to signal them that men are at last able to travel this far from earth; and the astronauts sit down to await the beings who will respond to the signal. A neat little open-ended thriller.

The screenplay by Kubrick and Clarke begins with a prologue four million years ago in which, among other things, one of those slabs is set up on earth. Then—with another set of characters, of course—it jumps to the year 2001. Pan Am is running a regular service to the moon with a way stop at an orbiting space station, and on the moon a similar slab has been discovered, which the U.S. is keeping secret from the Russians. (We are never told why.) Then we get the third part, with still another set of characters: a huge spaceship is sent to Jupiter to find the source or target of the slab's radio waves.

On this Jupiter trip there are only two astronauts. Conscious ones, that is. Three others—as in *Planet of the Apes*—are in suspended animation under glass. Kubrick had to fill in his lengthy trip with some sort of action, so he devised a conflict between the two men and the giant computer on the ship. It is not exactly fresh science fiction to endow a machine with a personality and voice, but Kubrick wrings the last drop out of this conflict because *something* has to happen during the voyage. None of this man-versus-machine rivalry has anything to do with the main story, but it goes on so long that by the time we return to the main story, the ending feels appended. It states one of Clarke's favorite themes—that, compared with life elsewhere, man is only a child; but this theme, presumably the point of the whole long picture, is sloughed off.

2001 tells us, perhaps, what space travel will be like, but it does so with almost none of the wit of *Dr. Strangelove* or *Lolita* and with little of the editing acuity of *Paths of Glory* or *Spartacus*. What is most shocking is that Kubrick's sense of narrative is so feeble. Take the very opening (embarrassingly labeled "The Dawn of Man"). Great Cinerama landscapes of desert are plunked down in front of us, each shot held too long, with no sense of rhythm or relation. Then we see an elaborate, extremely slow charade enacted by two groups of ape men, fighting over a waterhole. Not interwoven with this but clumsily inserted is the discovery of one of those black slabs by some of the ape men. Then one ape man learns that he can use a bone as a weapon, pulverizes an enemy, tosses the weapon triumphantly in the air . . . and it dissolves into a spaceship thirty-three years from now. Already we are painfully aware that this is not the Kubrick we knew. The sharp edge, the selective intelligence, the personal mark of his best work seem swamped in a Superproduction aimed at hard-ticket theaters. This prologue is just a tedious basketful of mixed materials dumped in our laps for future reference. What's worse, we don't need it. Nothing in the rest of the film depends on it.

Without that heavy and homiletic prologue, we would at least open with the best moments of the film—real Kubrick. We are in space—

immense blue and ghastly lunar light—and the first time we see it, it's exciting to think that men are there. A spaceship is about to dock in a spaceport that rotates as it orbits the earth. All these vasty motions in space are accompanied by "The Blue Danube," loud and stereophonic on the sound track. As the waltz continues, we go inside the spaceship. It is like a superjet cabin, with a discreet electric sign announcing Weightless Condition with the gentility of a seat belt sign. To prove the condition, a ballpoint pen floats in the air next to a dozing passenger, a U.S. envoy. In comes a hostess wearing Pan Am Grip Shoes to keep her from floating—and also wearing that same hostess smile which hasn't changed since 1968. When the ship docks and we enter the spaceport, there is a Howard Johnson, a Hilton, and so on. For a minute our hopes are up. Kubrick has created the future with fantastic realism, we think, but he is not content with that, he is going to do something with it.

Not so. Very quickly we see that the gadgets are there for themselves, not for use in an artwork. We sense this as the envoy makes an utterly inane phone call back to earth just to show off the mechanism. We sense it further through the poor dialogue and acting, which make the story only a trite setting for a series of exhibits from Expo '01. There is a scene between the envoy and some Russians that would disgrace late-night TV. There is a scene with the envoy and U.S. officials in secret conference that is even worse. I kept hoping that the director of the War Room sequence in *Dr. Strangelove* was putting me on; but he wasn't. He was so in love with his gadgets and special effects, so impatient to get to them, that he seems to have cared very little about what his actors said and did. There are only forty-three minutes of dialogue in this long film, which wouldn't matter in itself except that those forty-three minutes are pretty thoroughly banal.

He contrives some startling effects. For instance, on the Jupiter trip, one of the astronauts (Keir Dullea) returns to the ship from a small auxiliary capsule used for making exterior repairs on the craft. He doesn't have his helmet with him and has to blow himself in through an airlock (a scene suggested by another Clarke story, "Take a Deep Breath"). Kubrick doesn't cut away: he blows Dullea right at the camera. The detail work throughout is painstaking. For instance, we frequently see the astronauts at their controls reading an instrument panel that contains about a dozen small screens. On each of those screens flows a series of equations, diagrams, and signals. I suppose that each of those smaller screens needed a separate roll of film, projected from behind. Multiply the number of small instrument panel screens by the number of scenes in which we see instrument panels, and you get the number of small films of mathematical symbols that had to be prepared. And that is only one incidental part of the mechanical fireworks.

But all for what? To make a film that is so dull, it even dulls our interest in the technical ingenuity for the sake of which Kubrick has allowed it to become dull. He is so infatuated with technology—of film and of the future—that it has numbed his formerly keen feeling for attention span. The first few moments that we watch an astronaut jogging around the capsule for exercise—really *around* the tubular interior, up one side, across the top, and down the other side to the floor—it's amusing. An earlier Kubrick would have stopped while it was still amusing. The same is true of an episode with the repair capsule, which could easily have been condensed—and which is subsequently repeated without even much condensation of the first episode. High marks for Kubrick the special-effects man; but where was Kubrick the director?

His film has one special effect that certainly he did not intend. He has clarified for me why I dislike the idea of space exploration. A few weeks ago Louis J. Halle wrote in *The New Republic* that he favors space exploration because:

Life, as we know it within the terms of our earthly prison, makes no ultimate sense that we can discover; but I cannot, myself, escape the conviction that, in terms of a larger knowledge than is accessible to us today, it does make such sense.

I disbelieve in this sophomoric definition of "sense," but anyway Halle's argument disproves itself. Man's knowledge of his world has been increasing, but life has, in Halle's terms, made less and less sense. Why should further expansion of physical knowledge make life more sensible? Still it is not on philosophic ground that I dislike space exploration, nor even on the valid practical ground that the money and the skills are more urgently needed on earth. Kubrick dramatizes a more physical and personal objection for me.

Space, as he shows us, is thrillingly immense, but, as he also shows us, men out there are imprisoned, have *less* space than on earth. The largest expanse in which men can look and live like men is his spaceport, which is rather like spending many billions and many years so that we can travel millions of miles to a celestial Kennedy Airport. Everywhere outside the spaceport, men are constricted and dehumanized. They cannot move without cumbersome suits and helmets. They have to hibernate in glass coffins. The food they eat is processed into sanitized swill. Admittedly the interior of Kubrick's spaceship is not greatly different from that of a jetliner, but at least planes go from one human environment to another. No argument that I have read for the existence of life elsewhere has maintained that other planets would be suitable for men. Imagine zooming millions of miles—all those tiresome enclosed days, even weeks—in order to live inside a space suit.

Kubrick makes the paradox graphic. Space only *seems* large. For human beings, it is confining. That is why, despite the size of the starry firmament, the idea of space travel gives me claustrophobia.

POSTSCRIPT. Kubrick cut nineteen minutes out of *2001* after it opened. I saw the film again and thought that the cuts did little to help the sagging, although the fact that they had been made at all contravenes those admirers of the picture who say that Kubrick was not concerned with such matters as action and suspense.

Those admirers certainly exist, as I got ample proof. Usually letters that disagree with my reviews do so in pretty angry and direct terms. I got a number of such letters about *2001,* but I also got a quite unusual response: about two dozen very long letters, from four to eight typewritten pages, calmly disagreeing, generally sad but generally hopeful that I would eventually see the light. They came from widely scattered parts of the country, from students, a lawyer, a clergyman, a professor, and others. Most of those letters must have taken their authors a full day to compose and to type, and I felt that this disinterested, quite private support (none of the letters was sent for publication) was the best compliment that Kubrick could have been paid.

But he received plenty of public support as well. There were long articles explaining the psychedelic base of *2001;* there were new systems of film esthetics, by erudite and articulate critics, that used it as a foundation. With many of these critics I admired the inner consistency of their arguments, but for the most part, I could not see much connection between the criticism and the film. For instance, one of them said that *2001* had aroused adverse criticism (presumably from people like me) because it overlooks "assumptions promoted by a certain kind of literary humanism," because "its politics are unnamable," and because "it presents a complex and sometimes exalting image of that technology which we've been told again and again is inhuman and, therefore, the enemy of both art and the human spirit."

How I would like to see that film! (In spite of what I said about the dehumanizing effects of space travel, it is obviously a wonderful pictorial subject.) But has this critic described *2001?* The entire prologue is a heavy cautionary tale about the animalistic moral inheritance of human beings. The entire closing, on Jupiter, busies itself in tying up the *humanistic story,* after the long detour on the expedition into attenuated visual effects. The politics, far from being unnamable, are explicit cold war stuff, made ridiculous *sub specie aeternitatis* and therefore all the more a satiric statement of humanist concern about the human spirit.

My conviction remains: that *2001* started as a "true" Kubrick film on themes to be found in Arthur C. Clarke's previous fiction; that, en route through the years to completion, Kubrick fell in love with his technical ingenuity and equipment, and dallied with it. For me, *2001* is the luckiest film since *Tom Jones*—even luckier, because it not only made money, its shortcomings (in my view) fit perfectly the needs of a school of contemporary estheticians, who made the most of their opportunity.

One last point. Some have said that this picture cannot be truly appreciated unless one is high on pot. I assume that pot might make it more enjoyable, but then pot would also improve *Dr. Dolittle.*

The Odd Couple

(May 25, 1968)

ROUGHLY speaking, there are two kinds of film stars: those who are what we yearn to be and those who represent us realistically. The first are impossibly attractive, and they embody the beauty inside us that the world never sees—from Valentino and Garbo and Dietrich to both Hepburns (Katharine and Audrey), Peter O'Toole, and Cary Grant. The second embody the good humor, honesty, and warm humanity that, we are all sure, are our hallmarks to the world— Mary Pickford, Jean Arthur, Shirley MacLaine, Jean Gabin, James Cagney, Julie Andrews, Spencer Tracy for a random sample. There are also hybrids, combining glamour looks and earthiness—Fredric March, Sophia Loren, Tony Curtis, Lucille Ball are a few—but it is the consciousness of the two original elements in them that makes the compound effective.

Now we have a new star of the second, everyday type, so phenomenally commonplace that he almost makes the commonplace magical—Walter Matthau. (Not a new actor, a relatively new *star*.) Every office-working male in the U.S. either looks and sounds like Matthau or has felt, at some time or other, that he did. Matthau personifies all our grouches and resignations and small, beer-can pleasures that compensate for the fate of being alive in the twentieth century, and all with a sour humor that fulfills our fantasies ("Wish I'd said that") and massages our repressions.

Also, he has sex appeal. I saw his latest film, *The Odd Couple*, amidst a morning audience of middle-aged ladies, and older. The atmosphere round about me suggested that he reminded them of all those thousands of nights in marriage beds on the upper West Side or in Great Neck or Shaker Heights, not as it would have been with Richard Burton but full of friendliness, of a familiar passion somehow enhanced by mutual remembrance of squabbles in the car on hot Sundays and trouble with the kids and meeting all those endless bills.

In this new film Matthau co-stars with Jack Lemmon, who is one of the hybrid star types—Joe Average but a bit too good looking and adroit to be only J. A. and latterly too obtrusive a *performer* to let his J. A. personality work.

The Odd Couple was adapted for the screen by Neil Simon from his Broadway success, and once again a play suffers from screen dispersal. Everything that is forced to take place outside the one original setting—a living room—suffers for it.

The basic gimmick concerns a divorcé (Matthau) who has allowed his large apartment to become slovenly, and a poker crony (Lemmon), freshly separated from his wife, whom Matthau invites to move in. Lemmon is a compulsive cleaner and housekeeper and soon begins, with his neatness, to drive Matthau batty.

For me, the best feature of this comedy is that it ignores completely the theme of latent homosexuality. It would have been irritating if the theme had been treated superficially; besides, the purposes of the play are quite adequately served by its surface motions without any shallow delving. The next best feature is that it has a lot of very funny lines—gags, really, produced by Broadway machinery *in excelsis,* still undeniably funny.

But, as Simon showed in *Barefoot in the Park,* he simply can't go the distance. He can't build a full-length play. *The Odd Couple* begins with humor based on an authentic situation and degenerates into mechanical horseplay that is interchangeable with lots of other Broadway comedies. Simon, who was originally a TV sketch writer, is still a sketch writer: he has an eye for an amusing situation and a gift of the gag, but his whole instinct is to finish fast. When the curtain rises *again*—on Act Two—we can almost hear him saying, "You mean there's more? What do I do now?" (His new hit, *Plaza Suite,* which I haven't seen, consists of three short plays.) These short-winded shortcomings are equally obvious in the film versions of his plays.

Gene Saks, who also directed the film of *Barefoot in the Park,* has directed *The Odd Couple* less laboriously but without distinction. The Technicolor is, in two senses, ghastly.

Belle de Jour

(May 25, 1968)

If Luis Buñuel's new film were by an unknown director and thus were not escorted by the usual phalanx of Buñuel panegyrists, it might be easier to see it as a moderately amusing restatement of a familiar theme, fairly well performed, frequently titillating. True, some of the editing is sloppy (like the bit where we follow Michel Piccoli from the ski lodge for no reason whatsoever); true, the symbolism is heavy (like Jean Sorel's glimpse of the wheelchair that prefigures his accident) and the flashbacks that explain the heroine's psyche are simplistic; true, Buñuel is overfond of opening a sequence with a shot of a set and letting the actors walk into it; and true, too, that the heroine's fantasy sequences are prosaically conceived. But the color is good, the pace is satisfying, and the brothel atmosphere is something that is always interesting to both sexes.

For *Belle de Jour,* Buñuel returns to Paris, where he started his film career in 1926. The basic idea of the picture was old even before the Empress Messalina painted her nipples gold and went out to stand on Roman street corners. The screenplay was adapted by Buñuel and Jean-Claude Carriere from a novel by Joseph Kessel, which Buñuel is said to dislike. It has some resemblance to the Giles Cooper–Edward Albee play *Everything in the Garden,* which was seen, with fit brevity, on Broadway this season—it concerns a respectable wife who works secretly as a prostitute in the afternoons. The Cooper-Albee play, not to elevate it one whit, tried to use the idea as an attack on the mendacities of bourgeois life. Buñuel, who has often animadverted against bourgeois society, uses the idea solely in subjective, psychopathic terms. A young woman, who is frigid because of childhood conditioning, has married a weak, gentle young man. She indulges in fantasies of flagellation and violation and, in her own bed, she recoils from her too kind husband. Through a series of clues dropped by friends, she finds a genteel expensive brothel, where, after some initial hesitation, she throws herself heartily into her work—always leaving at five so that she can be home before hubby. We see a succession of odd patrons who arouse her; but she becomes thoroughly infatuated with the toughest of the lot, a brutal gold-toothed young thug. Her afternoon activities eventually lead to disaster for her poor husband, to her retirement from her new profession, and to her return to the fantasy life with which she began.

Her story is unrelated (as far as we are shown) to social-cultural causes, nor is it an excursion into Sadean freedom. Except when he is dealing with rape, Sade usually writes of people who have full sex lives of one kind or another, or of virgins who are eager to begin, who extend their sex lives in order to extend the borders of experience and to rebel privately against custom and religious proscription. Buñuel's heroine is not a virgin yet she has, figuratively, no sex life; her expedition into whoredom is psychologically compelled, a drastic means to redress early emotional injury rather than a "normal" person's deliberately expanded freedom.

There is a passage from Friedrich Engels about novelists that Buñuel, in a 1953 lecture, applied to film makers:

> The novelist will have acquitted himself honorably of his task when, by means of an accurate portrait of authentic social relations, he will have destroyed the conventional view of the nature of those relations, shattered the optimism of the bourgeois world, and forced the reader to question the permanency of the prevailing order, and this even if the author does not offer us any solutions, even if he does not clearly take sides.

In *Belle de Jour* the text is Krafft-Ebing, not Engels. By Engels' standards, since Buñuel has dealt with a sickness that could occur in *any* society, the film maker has neither shattered the optimism of the bourgeois world nor forced us to question the prevailing order, and has not in this instance "acquitted himself honorably." Instead of following the guidelines he himself has set, Buñuel has merely provided a diversion for the bourgeoisie—a very mild diversion at that.

His cast includes two men who have acted for him before, Michel Piccoli, as a rich hedonist, and Francisco Rabal, as an older thug. Along with Piccoli, there are two other actors from another recent piece of French near pornography, *Benjamin*. (A coincidence?) Pierre Clementi is the young tough, and Catherine Deneuve plays the title role. Clementi is much more credible here in ugliness than he was as the golden Benjamin. Miss Deneuve, who incidentally provides us with an Yves Saint Laurent fashion show, looks dainty in her underwear, but her face is not much more expressive than her navel.

A Face of War

(June 1, 1968)

DOCUMENTARIES—it is not superfluous to note—are supposed to document. Often, in these days of cinema equivocation called *cinéma vérité,* documentaries consist of pseudo-impromptu interviews, carefully steered and tendentiously edited. Obviously there is no such thing as utterly objective reporting, but there is such a thing as objective intent. Propaganda films are often desirable; subjective nonfiction films can be good art; but a documentary might be defined as fact dramatized but not distorted by either prejudice or zeal.

A Face of War is one of the best documentaries I have seen on its subject. Eugene S. Jones, an unknown name to me, is its principal author. With two other cameramen, J. Baxter Peters and Christopher Sargent, with a sound recordist named Robert Peck, Jones spent ninety-seven days in 1966 with Mike Company, Third Battalion, Seventh Marine Regiment, in Vietnam. (More than half of Mike Company were killed or wounded in those ninety-seven days.) The result of their work—as edited superlatively by Jono Roberts—must stand as a lasting record of the American military experience in Vietnam and of the foot soldier's life today—which is to say, the life of the Assyrian or Babylonian foot soldier as modified by modern technology and social attitudes (including his own) toward the soldier's occupation.

The first fact documented by this feature-length film is that today's soldier is a packhorse. Surely no fighter has had to *carry* so much into battle since the medieval knight. After that, as a spectrum of experience, the film's range is wide, yet not shallow. From the first shot (a rifle being loaded) to the last (men trudging into twilight), there is a feeling of multifold empathy and many simultaneous realities. The camera falls to the ground when men throw themselves down for safety; it fights through the helicopter wind as men carry the wounded to be evacuated; it scans soldiers' faces in a native hut as they watch a woman give birth; it twists with men in another kind of agony; it follows the string wire of a booby trap being disarmed; it attends mass and attends briefings; it huddles around a small campfire.

To itemize details is possibly to make you think you have seen it all before on TV and in other documentaries. The triumph of Jones's film is that, in great measure, you are right, and yet it is still extraordinary. Jones and his colleagues are better photographers, their sound track is more vivid, their film as a whole is more perceptive of the *fate* of everyone in it—marine and VC and civilian—than any TV or other filmed report I have seen about Vietnam. Besides the superior treatment of the expected, there are some unique touches.

For example, when we see some huts burning, we get the usual faces (yes, *usual!*) of women watching and weeping; but Jones also shows us U.S. soldiers watching *them:* stilled, uncertain, almost impatient at the interference with their feelings.

Further, there is the fine editing by Roberts. So much of *A Face of War* is excellently woven, in flow from shot to shot and in the shaping of whole sequences, that (in the best sense) it takes on the quality of a fiction film for which the shots were manufactured. In one sequence, we pan with the muzzle of a flame-throwing tank gun from left to right, then cut to a reverse movement by a similar gun, then cut to a reverse of *that,* and so on, at slightly increasing tempo. The movement of these horrible weapons is graceful; by editing to reveal the grace, Roberts has emphasized the horror.

This is an apolitical picture, unlike some other recent Vietnam films. It will not (nor was it intended to) alter anyone's views on the necessity of the war, whatever those views may be. It simply attempts to crystallize the experience of being *in* it: from the fighting to a football game (during a lull) in the mud to Hanoi Hannah broadcasting a Guy Lombardo record of "Jingle Bells" to the conversations about home. A sergeant, kneeling next to a wounded Vietnamese, says, "I'll never get used to writing these casualty tags for civilians." There is more pith in that remark—in its very spontaneity—than in all of *Soldiers,* Rolf Hochhuth's new elephantine play on the same subject.

Les Carabiniers

(June 1, 1968)

JEAN-LUC Godard made *Les Carabiniers,* his fifth feature, in 1963, and was evidently aware from the start that there was no point in making a fiction film about the horror of war simply to convey the horror. By 1963, everyone who would see his film would be well aware of the horror—partly because of documentaries. The only point in making fiction about war was to go beneath the surface of slaughter, to stain—in the lab sense—some virus strains in human behavior.

Godard chose to make his film in a childlike way, with few actors, minimum verisimilitude, fabulist simplicity, and home-movie techniques. (Which must have entailed much restraint, since his cinematographer was the wonderful Raoul Coutard.) The whole film, except for some inserted newsreel shots of carnage, is conscious playacting. Nobody is frightened of killing or dying. Condemned prisoners walk to the wall, knowing that *we* know that they will get up again after the camera moves away, that there are blanks in the Sten guns. Symbolic ballet—not gripping realism—is the mode.

Godard made this film in midwinter, with bare hard ground and gray light; we sometimes see the steamy breath of the actors. Two young men and two girls live in a shack in the middle of an immense ugly field near a town (unnamed, as is the country). Two soldiers drive up one day in a jeep, wearing fictitious uniforms, to deliver notices from the "King," telling the young men that they must come along and be *carabiniers* (riflemen) in the King's war. The young men are neither bright nor appealing (a good touch, their membership in the commonalty), and they are quickly won over by descriptions of loot they will garner and by advance permission to break children's arms. They go along with the soldiers. The girls soon entertain a gentleman caller.

Postcards are a main device. Many Brechtian titles are flashed on the screen, describing the young men's war experiences, all of them in postcard scrawl (we frequently see the girls going to the mailbox to pick up the cards); the bulk of the film consists of scenes showing what the postcards have already described. And when the young men come home, their loot consists only of a valise full of postcards, which —again in a Brechtian-gull way—they treat as payment for their years of wandering and killing. There is even an organ tune reminiscent of Weill under the very long scene where they pull out the cards, one by one, with feebleminded pleasure as if they had indeed brought home the palaces and treasures and women that are pictured. Then there is a postwar revolution. The war has been lost; the royalists accuse democrats, Marxists, and Jews of a stab in the back; and in the hubbub, when the two young men go to collect some real loot, they are shot.

To emphasize the primitivist effect, Godard uses some early film references. Roy Armes has noted, in *French Cinema Since 1946,* that some sequences pointedly refer to films by Louis Lumière, the

French film pioneer. There is a scene in a movie theater that suggests Keaton and Sennett comedies. There are several suggestions of Chaplin's *Shoulder Arms*. Also, Godard sometimes quickly fades to black, then quickly fades in on the same shot—a stylistic equivalent of the jump-cuts in *Breathless*—as if to make sure we never forget that we are watching the "pretense" of a film.

In the nine-year spate of Godard's work, *Les Carabiniers* seems to me one of his more satisfactory films. It is not necessary to construct a style *for* it, as some Godard enthusiasts have done in other instances, using the record of his cascading whims, energies, and impulses as proof of an esthetics—simply because all these things are photographed and permanent. *Les Carabiniers* has shape and commitment. (One serious flaw is the use of newsreel shots: real bodies diminish the larger, unrealistic points that the film is trying to make.) But, leaving aside the tedium of such scenes as the long postcard inventory, there is, I think, a barrier between this film and us.

Godard has—soundly—decided to view his story about the slaveries of war from an abstract angle. But he then proceeds to tell us nothing we don't already know—"know" in the emotional or psychical sense. Imagination is not enlarged for us; cognition is not deepened. We simply watch a charade which, with some exceptions, is largely foreseeable once it has begun. With so little of new depth or powerful restatement, the film begins to backfire on Godard. Because he uses his method to relatively weak ends, the film eventually seems only an advertisement of his cleverness in choosing the method.

Rosemary's Baby

(June 15, 1968)

SOMEWHERE near the middle of *Rosemary's Baby,* three things begin to happen, and the second half of the film becomes highly effective. First, the cumbersome building-block method at the start is abandoned for the *use* of what has been built. Second, Mia Farrow begins to justify her presence as Rosemary; we see that she has been cast with a view to what happens ultimately, rather than for initial (failed) charm. Third, our expectations change toward the director, Roman Polanski. We realize that he must no longer be burdened with the standards he set for himself with *Knife in the Water;* since that picture, he has only been trying to entertain us and, on this level, he is at last succeeding.

The screenplay, based on Ira Levin's novel, is credited to Polanski. This is remarkable. Even though much of the dialogue comes from the book, even though Polanski's fluency in English has grown astonishingly, it is still a fact that he left Poland less than five years ago with no English at all.

The story, familiar to many, is about a young New York couple who get trapped in witchcraft, and it thrives on the contrast between Manhattan modernity and ancient magic. The setting is a famous old apartment house on Central Park West. A young actor makes a bargain with a band of witches—technically, a coven—in return for success in the theater that they will fix supernaturally for him. The deal obliges him to let Satan beget a child on his wife, who is quite unaware of the Faustian pact but who is to be a "black" Mary bearing a "black" Messiah. The plot focuses on her slow realization of what is happening in and around her during her pregnancy, her efforts to save her forthcoming child, and, after delivery, her acceptance— out of irresistible mother instinct—of her diabolic infant son.

Most films that begin feebly finish feebly. *Rosemary's Baby* is an exception. The opening pan along the New York skyline, then down the front of the apartment house, is trite. The first scenes are disappointing: the young couple and the renting agent, the couple and an old friend at dinner, the young wife and another girl in the basement laundromat. They lack the sense of control that even Polanski disappointments like *Repulsion* and *Cul-de-Sac* showed in every frame. Much of the shooting in the early part is from floor level, possibly to suggest overhanging dread even in bland scenes, but it doesn't work. For too long, our only interest is in an elderly couple next door. Sidney Blackmer's fruity acting is here overripened to suggest decadence. Ruth Gordon's stridency and gnomishness, which have usually distracted me in the theater, are here skillfully used to imply that her comic everydayness covers something baleful. Then, too, as the film begins to jell, Miss Farrow begins to hold us. Through sheer unattractiveness, as well as inadequacy, she flounders at first in what seems a hip comedy with mystery overtones. When she becomes the emaciated, trapped victim of a mystery with comic overtones, she gets much better.

Some of the other casting is questionable. As the young couple's older friend, Maurice Evans is an elocutionary dud. As Miss Farrow's husband, John Cassavetes has little flavor and only imitative sharpness—but, admittedly, this makes him resemble many of the people in the theater world that his character inhabits. Ralph Bellamy, as a bearded Jewish obstetrician, is not markedly different from Ralph Bellamy as Franklin D. Roosevelt, save for the whiskers.

The Technicolor camerawork of William Fraker is acceptable, except in the exteriors, where it is unacceptable. The process photography—dreams and so on—by Farciot Edouart is much more clearly "seen." The most interesting point about the music, which is too whooped up in the climactic scene, is that it was written by Christopher Komeda, who, as Krzysztof Komeda-Trzcinski, has written the scores for almost all of Polanski's films since they were students together in Poland.*

As for Polanski himself, he is teaching us how to regard him. *Knife in the Water,* his fine first film, was a tight little Sartrean engine of internal forces. Since then, the horrors in his films have become much more external, at best merely entertaining. *Repulsion,* a chronicle of psychotic murders, was coolly frightening, if largely gratuitous. *Cul-de-Sac* was a far-out thriller-rag, less successful but sometimes ingenious. *The Fearless Vampire Killers,* which Polanski says was mutilated by the distributors, was an amusing idea for a *Dracula* spoof, but it completely misfired. *Rosemary's Baby* seems to settle in right where he wants to live: as a manufacturer of intelligent thrillers, clever and insubstantial. Only a director with wit could have made the witchcraft credible. Only a director with real cinematic gifts could have made a sequence like the one where Rosemary barricades herself in the apartment or the childbirth scene. Only a director satisfied with ephemera could have lavished his gifts on the whole project.

* Komeda died, as the result of an accident, on April 23, 1969, aged 37.

Fist in His Pocket

(June 15, 1968)

THE young Italian Marco Bellocchio proved his talent with *China Is Near,* his second film, with which he made his American debut. Now his first film, *Fist in His Pocket,* has arrived. It is excellently acted, and Bellocchio has directed with a wonderful modern impatience—impatience with irrelevant detail, mechanical transition, and conventional sentiments. But all the talents—of cast and director —spill out of the leaky script. Written by Bellocchio, it is intended as dark domestic tragedy; it ends almost as melodramatic parody.

An upper-middle-class family in a north Italian villa. Winter. (It's not always bleak in northern Italy, but no young intellectual Italian director wants any suggestion of *"O Sole Mio"* in his work.) Mother, a widow, is fiftyish and blind. Four grown children: a lawyer (betrothed), an idle lovely girl, an epileptic son, and a demented son. The epileptic decides that, to free the eldest son, he will do away with all the others, including himself. A car crash on a particular trip would settle it, but he loses his nerve at the wholesale job and goes to work piecemeal. Subsequently he pushes Ma off a cliff and holds the loony under his bath water. When Sis discovers this (after a spot of incest between her and the epileptic), she is shocked into a backward fall downstairs, cracks her spine, and is paralyzed. It all ends with the sister bedridden, helpless to aid the epileptic in the next room, who is writhing on the floor while a phonograph blares *La Traviata.* Violetta's penultimate note in Act One is prolonged on the sound track to finish the film with a scream.

This synopsis does Bellocchio no injustice because what starts as Italianate Faulkner is soon so contrived in its dooms that it becomes a long gallery of grotesques, without much relation to us or to the demons in contemporary society. *China Is Near,* despite its dogged Godard imitations, is a much more relevant work. *Fist in His Pocket,* made a year earlier (1966), seems an overreaction to the sugar candy of most films. You can almost hear Bellocchio swearing to revenge himself for all the movie floss he had been forced to endure in his twenty-six years.

Yet it is finely executed. Three actors especially deserve mention.

Lou Castel, the epileptic, is more than credible in perversity. Marino Masè, the oldest son, has strength and depth. The daughter, Paola Pitagora, is not only a good actress, she has one of the loveliest Italian faces on film record.

Petulia

(June 29, 1968)

RICHARD Lester's *Petulia* is a dazzling smear across the screen—fuzzy, unrealized, stunning. He has taken some unlikely materials and, in his gifted hands, they remain unlikely materials; but much of what he does with them is wizardry.

The script by Lawrence B. Marcus, from a novel by John Haase, has some bright dialogue, kooky and kurrent, but it does not validate a basically sudsy, unsatisfactory drama. Petulia is an English girl married to the scion of a rich, reactionary San Francisco family. At a charity ball ("Shake for Highway Safety") she sees a fortyish divorced doctor and at once tells him, with leaden impishness, that she wants an affair with him. He is reluctantly acquiescent; they go to a motel where she gets (euphemistically) cold feet. Subsequently, however, she pursues him, and they bed. Her young vicious husband finds her in the doctor's apartment alone and beats her to the edge of death. After the doctor saves her life, her in-laws spirit her out of the hospital, and she rejoins her husband, of whom she is perversely fond. She refuses the doctor's rescue. They glimpse each other occasionally through the following months. At last she turns up at his hospital to give birth. Even at this moment, the doctor offers to rent a private ambulance and flee with her. She declines ruefully. They part, forever, with a quip.

The point, presumably, is interchange. Early on, the doctor tells a colleague that he got divorced only because he felt nothing in his marriage and wants "to feel something." Petulia feels excessively, and is unstable. At the end he says to her, "You've turned me into a nut"; she has presumably become stable. Unfortunately, this remains an equation on paper. He has not become a nut, as far as we can see, or a more feeling person than he was. The only serious change in Petulia is that at last she is satisfying her starved maternal urge, which is not the same, necessarily, as undergoing a change of character.

The last portion of the film concentrates on Petulia's reconciliation with her husband. Up to then the focus has been on the doctor, and Petulia has been a character in *his* story. This final shift is a tail that keeps wagging but never wags the dog. We are made to go through a simulated nine months—discursive and anticlimactic—all for the sake of that brief *ante partum* scene. Also, throughout the film a lot of her kookiness is drearily arch, including a long sequence about a tuba.

But even this same script would be much more enjoyable with some improved performances. Julie Christie, the Petulia, is a photographer's model, not an actress, so incompetent that she doesn't even seem beautiful any more. Her first film appearance, in *Billy Liar,* was her best—just as a striking face. John Schlesinger, who made that film, evidently became infatuated with her face and adored her, by camera, all through *Darling.* She floundered as Lara in *Dr. Zhivago.* Truffaut doubled her dullness by giving her two roles in *Fahrenheit 451.* Then Schlesinger cast her as Bathsheba in *Far from the Madding Crowd,* which exposed her as a mod Chelsea minitalent wallowing about in Wessex. (Nicholas Roeg, who photographed her well in that film, does it again here.) Now Lester has cast her in light brittle comedy, the kind of acting that is probably the most technically demanding—and she is a girl with almost zero technique! All she has are her face, a pleasant voice, and some weak ability to imitate actresses she has seen. And what has happened to her lower lip? It was always large; now it just seems to hang there, immobile, while she chatters behind it.

Richard Chamberlain, as her spoiled young husband, looks pretty enough but lacks the necessary hint of evil, of filth under silk. Joseph Cotten, his Birch-barking father, has not given a really satisfactory performance since *Citizen Kane.* Arthur Hill, who plays a colleague of the hero's, is a good actor but here is casually unintelligible much of the time.

With these people replaced or improved, results would have been much better, because the doctor is played by George C. Scott, who —though not the world's most lovable personality—is a tremendously appropriative actor. With the restraint that, happily, he has been developing, he takes possession of a role in a way that makes it unimaginable apart from him, which is at least one definition of talent. I'm unable to believe that Petulia sees this stranger across a crowded room and decides wham-bang that this is an enchanted evening; nevertheless the role and the film quickly become Scott's domain. (That is why the shift of focus away from him is tedious.) Pippa Scott is right as his mistress, the kind of pleasant woman whom no one ever marries. Shirley Knight, as Scott's ex-wife, stops—with brakes screaming—just this side of Sandy Dennis-type sensitivity. A scene between her and Scott in his apartment, full of sublimated desires and tensions, is one of the best in the film.

But all the above—and this is why the picture is remarkable—really tells you very little about *Petulia*. It could describe a well-intended Lumet or Frankenheimer fumble. The difference, the excitement, is visual, is Lester. He was disappointingly misguided here, I think, in his choice of material and players, but once in, he was *in*—uncorked and bubbling. He *attacks* his films, probably too frenetically but with such effervescence of ideas, such eagerness to do everything he can imagine, and with such incessant imagination, that his ebullience seems to spill out of the frame all over us. When we begin to think "Enough already!" he washes our objections away with another torrent of visual energy, and when we begin to object again, another torrent, and so on. It is something like being on a beach and getting knocked down by a wave every time you try to regain your feet. It may not be the best swim of your life, but there is a kind of beauty in the sheer power of the surf.

As with Lester's past work, *How I Won the War* and *The Knack* and the Beatles films, it is impossible to describe any one minute of *Petulia* in less than thousands of words. He works with a barrage of visual jokes, cross-references, contrapuntal strands, and mere mood inserts (like recurrent quick microscopic shots of mobile cells which suggest what we see with our eyes closed). I note a few indicative items. Through the main story he sifts in another story, telling us how Petulia brought back a Mexican boy from Tijuana. This understory is breached in subliminal flashes, puzzling at first, until we see the relevance to Petulia's child-hunger. The first flashes are bewildering, as if we had stumbled on irrelevant privacies, but Lester is making a mental model for us: showing us how strangers think things unknown to us and how, as we get to know them, we know somewhat more of what they are thinking.

San Francisco is distilled, in this Technicolor film, almost as knowingly and humorously as London in *Blow-Up*. The Japanese garden in the large city park; cable cars, of course, but *very* briefly; rows of tract houses that stretch over hills like tombs in a giant graveyard. A blue-filter scene in a supermarket dawn evokes chilled inhumanity among the frozen foods. Background detail is touched in lightly, to make the city and the world: a pair of leathery lesbians drift across the fringe of a few scenes; the PA system in the hospital calls—among the sick—for a doctor to move his parked car; a puzzled hostess reveals her manipulated life in a glance when, from her dining room in the distance, she watches Scott leave her husband's home suddenly. Lester works by a system of cinematic mosaic, a lot of swift whirling pieces that, through careful selection and their very number, resolve into pattern and hard surface.

Much has been said about Lester's relation to TV-commercial style. In London (where he usually operates) there is a small club-restaurant off St. James's Street frequented by film directors where, I'm told, they meet and try to decide by details of style which of them made the commercials that were seen on the telly the previous night. These men include, or did for many years, the leading directors in England. Commercials have fed many of them literally and fed Lester in two senses. He uses the jam-packed clevernesses of our time to comment on the clevernesses of our time; to catch—to exploit seriously—the audience whose eyes and ears are whetted to those clevernesses. (Specialists say that the commercials are the most popular TV items with children. What will films be like twenty years from now?) He understands the psychology, the media orientations behind the clevernesses, and they are what interest him. To say that his pictures are only extended TV commercials is (in large analogy) something like saying that David Smith's sculpture is only aggrandized arc welding.

My own reservations about the subliminal barrage are somewhat different. It seems to me that the quick-flash reference sometimes hinders imagination, instead of helping it: deprives both the actor and the audience of some rightful imaginative exercise. Everything is *shown*. (If Petulia had a line about heaving a rock through a window to get the tuba, and if Miss Christie could play it, it would be funnier than showing the incident.) But the idea of tracing out the lightning of the mind on film is relatively new; needs exploration; and Lester is one of the best directors exploring it.

Here he has spent his talents on inferior materials and performers. But at least *Petulia* frequently scintillates where, in lesser hands, it would have been pretty dull.

Inadmissible Evidence

(July 13, 1968)

IF there is one thing that John Osborne can do, it's write dialogue. This is a handy knack in playwrights—and not to be taken for granted even among famous ones. His play *Inadmissible Evidence* has faults, but one of them certainly is not pallid, clumsy, incredible, or trite dialogue. In speech, at least, he is proof that in England the line of John Marston is not dead yet. Osborne's particular brand of acid-torch rhetoric is the *body* of this play. He creates characters, sprinkles humor, and tells something of a story, but it is in the very texture of his language that the true drama resides. It is in the words blue-slashing through the air that he fixes the modern England that he would like with all his soul to loathe, if only he could.

I saw *Inadmissible Evidence* in London and New York and each time was struck strongly with several matters. Here are some of them. The nightmare prologue, which is virtually a long monologue for Maitland, the solicitor hero, is really the play in capsule. It could be played separately and would give us the essence of Maitland and of Osborne's views on most of what follows; mainly, the rest of the play is varyingly successful articulation of what is compressed in the capsule. The mind of Maitland is really the arena of the play; everything else exists there or else is significantly excluded. The character of Maitland—embittered, self-hating, viciously funny, lecherous, lost, agonizingly self-aware—is a plateau, not a cumulative growth. The exposition of that character goes on too long, but the wonder is that Osborne can find as much variety and nuance in it as he does. There is some tension or change going on between Maitland and almost every other character, yet the sense we get is not of dramatic progress but of dissection. The short scenes in which the clients appear and tell their troubles are excellent—lives not merely summarized but distilled. The very existence of the play itself is a statement of social relevance: it is only eight years from *Look Back in Anger* to *Inadmissible Evidence,* and Osborne has shifted in subject from the outsider looking in to the insider looking wistfully out.

Whatever it lacks in cumulation and growth, the play is committed to its language and to its theater *form.* This commitment is what is lacking in the film. Where the play had courage of conviction in its form, the screenplay—by Osborne—settles for imitating the form of many other films. The opening monologue is drastically condensed and partially replaced by (dream) sequences of Maitland en route to the dock; we are shown his home and his commuting; we see him frisking on the beach with his secretary; we see Maples the homosexual being arrested in the subway lavatory, and so on. All this seems to me meek subservience to clichés of what makes films "visual." Osborne would simply snort if anyone tried to impose equivalent formulas on him in the theater. The play need not merely have been reproduced on film (although Peter Brook's *Marat/Sade* film showed that this—or nearly this—can be cinematically practical). But most of the cuts in the play weaken the film, and most of the expeditions simply state prosaically what had been vividly suggested. Two "improvements" are particularly poor: the walks along ugly streets with ugly faces, to show how debased modern life is, are an unhelpful bow to early Tony Richardson; and—yet again—the strip-club scene. Britain now should ban scenes in strip-clubs to symbolize failed depravity.

What is left of the play is often very pleasantly vitriolic and sometimes quite moving. Anthony Page, who directed both play and film, has again done well by his actors; his camerawork lacks the immediate sense of self-assurance, of confident individuality, that his stage work had. Nicol Williamson presents a large portion of his original Maitland, surely one of the prodigious acting performances of the decade. (But the film diminishes him in a sense, besides condensing the role. In the theater he earned our added admiration for a marathon sustained effort; here we know it was all done in bits and pieces.) Isobel Dean as a divorcing woman is beautiful in her brief scene, but the script omits her character's fantasy reappearances that, in the play, bore out the opening fantasy mode. Eileen Atkins is fine as the disgusted mistress-secretary. There is one advantage in the perambulating film script; it gives us more of a chance to see Jill Bennett, the "permanent" mistress. She is warm, strong, but tender, just the kind of mistress that any right-thinking man would want and that a wry-thinking man like Maitland desperately needs.

Osborne and Page have missed a chance in this film, I think. The original *Inadmissible Evidence* was too long for what it had to say—the evidence was rather too admissible—but it had its own sense of being, which is here badly cracked. A play that was conceived as an increasingly bad dream has been made into a grittily detailed, naturalistic film. The energy that went into making the picture "visual" would have been much better spent on developing the dream concept, rather than on pavements. The central image of the play—Maitland trapped in his office, sinking in a swirling sea, sucking for life at the end of a telephone—is lost.

For Love of Ivy

For Love of Ivy, which stars Sidney Poitier, is based on an original story by Poitier. Original, that is, in the venerable Hollywood sense: "Let's make a new picture right down the middle of the old alley." Poitier is a big-shot gambler with a heart of gold who runs an honest business as a front. (Clark Gable? Spencer Tracy?) Two rich teenagers, brother and sister, cook up a plot to involve him with their family's maid (Claudette Colbert? Irene Dunne?) so that, with male attention to keep her from being restless, she'll be content to remain in their family's Long Island home. The gambler has to be blackmailed into meeting the maid; she is equally reluctant to meet him. Of course the pair are destined to fall in love. The only missing step in the formula is Loathing at First Sight. Instead we get here an immediate statement from both that they have no intention of marrying—each other or anyone else.

Naturally the formula has been given plenty of contemporary data.

There are hip references and swinging dialogue. There is that bed scene which is now par for the intercourse. And at the end, no misty march to the altar, just a modish mutual resolve to stick together and see how things work out. But all the above is only new chromium on a standard model. *Ivy* is one more bonbon from never-never land except—and the exception is the whole point—this time the lovers are black.

Already there has been some stiff comment about this film's betrayal of the principles of black liberation, about its subscription to bourgeois ideals and fantasies. Some of this comment has come from people who applauded Poitier's last picture, *Guess Who's Coming to Dinner,* a Stanley Kramer opus that rubbed long-lasting 5 Day deodorant all over that nasty race problem so that it could sit next to us at table. In my view *Ivy* is more honest and certainly more skillful. Unlike Kramer's picture, it does not temporize fantastically with grave issues, it opts completely for the fantastic. Under the well-spun floss it seems to be saying: "You want integration? OK, we're going to integrate you out of your minds. We're going to integrate with white America's heritage of twentieth-century pop myths."

The only valid question for this film is: Does it use its myths well? Are we given two engaging vicars in a romance that we can vicariously enjoy? The answer is yes.

I missed, deliberately, Poitier's three other recent romances (of differing kinds): *Lilies of the Field, A Patch of Blue,* and *To Sir with Love.* Nothing I heard or read about them persuaded me that they were my brand of cop-out. I went to *For Love of Ivy* principally because the title role is played by Abbey Lincoln, who was quietly charming three years ago in *Nothing But a Man* (a far superior film). She is even more charming here.

As an actress Miss Lincoln is competent enough, but I would not be surprised to learn that, where the role exceeded her ability, it was tailored back to fit. She is not conventionally pretty, although she has lovely eyes. What she has is an irresistible personality: tender, vulnerable, humorous, and endearing. If she gets enough good parts that lie within range of her dignity and truth, I see no reason why she should not be the first big female Negro film star. There may not be much more to her than the convictions of her person, but they are enough to make her worth watching, pleasurably, for some time to come.

Poitier leveled off as an actor quite a while ago, but it is a good level, always forceful and credible. As a star, he has entered into the tacit sex dialogue that is a requisite between a star and his audience. I saw the film at a matinee attended by lots of ladies of both colors and all ages. When Poitier stepped out of his car to make his entrance, a sigh went around the theater, an embrace of welcome. When Miss Lincoln visited his *Playboy*-type pad and he appeared in a V-neck pullover with no shirt beneath, there was a puzzled but expectant sigh. ("Why so unchic, Sidney? But of course you must have your reasons. Forgive us for doubting. We'll wait and see.") In the bed scene, where he seemed to be naked, I had to switch off my audience radar. Too lurid.

Daniel Mann directed, and his outstanding quality here is his reliance on his stars. For instance, in the inevitable ultimate reconciliation after the inevitable penultimate quarrel, there is a long scene in which Poitier kneels next to Miss Lincoln's chair and talks her round. Mann very wisely just puts the camera on them, fairly close, leaves it alone, and lets their personalities work, which they do.

The history of films is, in one aspect, the history of dangers and problems given ninety minutes of reprieve, whether runaway railway trains or impregnable class barriers or the heroine's blindness that is cured in time for the final fadeout. We all enjoy it when well done because, briefly, it endows the chaos of the adult world with the order of the nursery. Here the subject of race has been fed into smooth, expensive romance, signifying that the concern about it is now widespread enough to breed widespread appetite for brief respite *in both black and white* (I was in a mixed audience); and signifying, too, that at least some Negroes have now reached a position to exercise the franchises of genuinely mass fantasy. In *Ivy*, Cinderella has been updated, enough to make it topical, not enough to distort the power of the perennial myth. The result, whether we like the premise or not, is one of the most likable "movie" movies in the last few years.

Targets

(*August 31, 1968*)

PETER Bogdanovich is twenty-nine, the author of several brochures
for the Museum of Modern Art Film Library and of numerous articles
about film. He has now written, directed, and produced *Targets* and
also plays a substantial role in it. As a film, it's minor; as a phenomenon, significant. So far as I know, *Targets* is the first picture
made in Hollywood by an American critic of the *auteur* school. France
has had many new *auteur* directors in the last decade, but Bogdanovich is the first American *auteur* to appear in the city that is a particular heaven for *auteurs*. All those Hollywood elements of com-
merce and popularity-groveling that seem restrictive to many of us
have meant little to *auteurs*. Their chief concern is with the way a
director handles the material he chooses—or is assigned. Many of us
think of Hollywood as, in general, the home of hacks or of good men
hampered. Here is an intelligent, utterly hip young man who *chooses*
Hollywood. His action and his beliefs have nothing whatsoever to do
with that other group of young film makers, the Underground or
Free Cinema. They are anti-Hollywood. The *auteurs* are, in one sense,
the first pop artists and cultists, but with a difference: they can see a
fourth-rate melodrama and know it is fourth-rate as a melodrama at
the same time that they glory in the director's use of the camera and
his expertness in film mythology.

Bogdanovich's story comes out of movies, flutters between movies
and life, then goes back into movies. It weaves two chief strands. A
famous star of horror films—Boris Karloff more or less as himself—
wants to retire. A healthy young Californian goes off his rocker and
shoots a lot of people with high-powered rifles before he is caught.
The two men cross paths at the beginning, though only the killer
knows it, and then face each other at the end.

The contrast, of course, is between the fabricated horror of Karloff
movies and the quiet horror of life, between the gloomy castles in
the Karloff film clips and the sun-drenched, Mustang-mounted
murderer. (Reportedly, the use of Karloff and Karloff clips were conditions set by the financial backer.) The strands are counterpointed
throughout and finally join. In the last scene the mad youth, in a

drive-in, is almost literally *in* a Karloff movie. He sits behind the screen, pokes a hole in it, and fires at the audience. Real death reaches them through the beam of the projector.

This contrast may sound thin in description, but description is words. The story could exist only on film. It is the presence of Karloff, attended by thirty-five years of our memories, that is essential; no actor playing a Karloff type would have sufficed. The clips from Karloff films are this picture's mythic bona fides. To immerse *Targets* further in pertinent unrealities, Bogdanovich binds his picture closely to films and screens and viewing. The killer, lugging his bag of rifles to the top of an oil tank in order to shoot at the freeway, has to pass an abandoned studio back lot en route. He and his family sit watching a late-night TV show like communicants at a stupid and stupefying mass. Karloff and his young director in the story (played by Bogdanovich, the director of this film) watch a segment of an old Karloff film on TV and praise the director, Howard Hawks, a god in the *auteur* pantheon. There are several references to well-known films. When the killer types a note, the camera gives us a huge close-up of the letters *d, i, e* appearing on the paper, just as the letters *w, e, a, k* appeared in *Citizen Kane*. Karloff's Chinese secretary quotes an old Chinese proverb that is a Spanish proverb in *The Philadelphia Story*.

Purely filmic qualities in *Targets* are used in still another way. The production designer, Polly Platt, coauthor of the original story (Bogdanovich wrote the screenplay alone), has made the various settings carry considerable weight. Karloff's hotel suite feels luxuriously unfriendly. The killer's home looks like a colored-plastic, fully automated trap. In fact, the *look* of his San Fernando Valley is the only explanation of his behavior. The two or three mumbled lines about his troubled mental state are feeble. The real motivation is the choking material perfection of his home. And it is not a rich home; its antiseptic, supermarketed tidiness is the middle-class fate. For me, the film's best achievement is in conveying—or at least suggesting—the killer's suffocation, as he makes the daily rounds of colored paper towels, snazzy wall ovens, wall-to-wall carpeting; thinking that this Saran-wrapped neatness is what lies ahead of him every single day of his life; thinking: Is this *it?* Is this *all?* And then boiling sideways in a violent effort to escape from the mechanism of the air-conditioned clock, murdering the warders who are part of this bloodless perfection. An added nice touch is that, after he kills his first three people, in his home, he lugs the bodies away and covers the stains with towels. It is an echo of what seems to have driven him mad to begin with.

If I am reading more into this than Bogdanovich consciously intended, then he is reaping some benefits of the powerful suggestiveness of film, which he *does* use consciously. But there are very serious drawbacks in this almost single reliance on sheer filmic powers. It can lead to the hyperspecialist, the man who knows one thing very well and little else, no matter how relevant. To the *auteur* the world is one huge film-studio prop room, from which he can select bits that will photograph well, virtually regardless of other considerations. I don't mean, in this case, social responsibility. Bogdanovich's distributors have pinned on a silly foreword to make the film seem an argument for gun control; they are apprehensive about one little sniper film in a country full of gunfire television, and newspapers and magazines, a country in which a presidential candidate (Wallace) draws loud cheers by promising to run over any protester who lies in front of his car.

No, the primary faults of *Targets* would exist if both of the Kennedys and Dr. King and those people on the Texas campus were still alive. Bogdanovich just cooked up some clever ideas out of the backer's prerequisites and filmed them smoothly with little regard for credibility or conclusion. What is the point of this film, once the gimmicky juxtaposition has paled? Bogdanovich has little to tell us about this killer—as artist, not sociologist. He uses the Karloff character like a clumsy puppet. Why does Karloff suddenly want to retire? The chief reason would have applied long before—that the real horror in the world has made him feel silly. The other reason, that he is passé, is patently untrue. And what has this retirement theme, which takes a good deal of time, to do with the story? And why in heaven does Karloff advance bravely on the trapped and armed murderer at the end?

That advance, and the token wounds that he and his secretary get from the sniper who has been polishing off people hundreds of yards away, are Bogdanovich's deepest bows to Hollywood fantasy.

And doesn't acting count for anything, even with a "visual" director? Nancy Hsueh, Karloff's secretary, is simply incompetent. Karloff—as he says of himself!—is no longer able to play a "straight" scene. Bogdanovich caps his misperception of acting by casting himself as the young director, apparently in the belief that all one needs in front of a camera is the ability to be natural.

Another curious point. Only the scenes of reference—cinema history or social satire—are credible. Virtually all the others are phony: those between Karloff and the secretary, the secretary and the director, the killer and his wife. Bogdanovich reveals little ability as yet to create anything of his own. He can only show us what he has seen.

To argue for more than filmic content in films is taken by some as an argument for literary or theatrical film. But such new directors of the past decade as Bellocchio, Bertolucci, De Seta, Olmi, Jessua, de Broca, Lester, Teshigahara, and Nichols have shown that film can be truly film without being *only* film. Bogdanovich, however, has grown up in an esthetics that exalts manner over matter—no, it tells us fundamentally that manner is all: that Samuel Fuller's *Shock Corridor* and Nicholas Ray's *Party Girl* and Preminger's *The Cardinal* and Hitchcock's *The Birds* and Hawks's *Hatari!* are excellent art works because of the directors' styles, that objection to the tacky stories is misplaced because the film is not in the story but in "the relationship between the director and his material" (Gavin Millar). To me, this seems the equivalent of the theatrical legend about the great actor who could pulverize you by reading the telephone book. I have never had the luck to be thus pulverized, but the legend does not maintain that the *ideal* is to have great actors read telephone books. I am unconvinced that any of those directors is as good when (even because) he uses fourth-rate material as when he uses material and performers that satisfy other expectations in us as well.*

It's good to welcome a new director with Bogdanovich's affinity for film making, his cinematic eye and appetite, his eagerness to edit incisively—in short, a man with exceptional facility. But facility can be mere glibness in a director who has no self other than his skills and his "in" enthusiasms. As the expression of a self, of a whole man who lives in the world and uses film to convey that fact, *Targets* is weak: little more than a checklist of professional venerations. It is noteworthy mainly as the first slickly made feature by a young American who has got his education and his bearings, so to speak, at the movies.

* There is further comment on the *auteur* theory in the review of *Lola Montes,* page 160.

Hunger

(*September 7, 1968*)

IN the film of Knut Hamsun's *Hunger,* Per Oscarsson gives one of the most searing screen performances I can remember. Wonderful. But the film itself, in its treatment of the novel and in its direction, is not good enough for him or for Hamsun, whom he is serving. The effect is as if an actor were giving a full-bodied performance of a Shakespearean hero in a film of one of Lamb's *Tales from Shakespeare.*

Hamsun's novel, published in 1890, is a work of genius. Among its achievements, it takes the figure of the artist out of the Romantic era, where he had served as a symbol of new sensibility, of individualism wrenching loose from state and church and social orthodoxy, and plunges him into the consequences of his decision, thus opening up the twentieth century. I. B. Singer has pointed out the relationship between Hamsun's anonymous hero and Raskolnikov. (Can it be mere coincidence that a picture of Dostoevsky hangs on the editor's wall in the film?) Rereading the book, I was struck by the way it prefigures such modern fiction as *Tropic of Cancer* and the William Burroughs novels. (There is a fine new translation of *Hunger* by the poet Robert Bly, available in a Noonday paperback.)

Hunger was Hamsun's first published novel, and, of course, its title denotes and connotes. It is an autobiographical work, telling of his early days as a young countryman in Christiania (now Oslo) when for days he had nothing to eat, when he chewed on scraps of paper or wood shavings or bones begged from butchers. It is also the story of his hunger to insist on his self. His famished body feeds his fantasies, as he struggles to scribble short articles, invents jocular relationships with passing policemen, invents and then realizes a dream of a girl, and trudges, trudges, trudges around the busy, cozy, wintry city. Possibly its supreme achievement is that the hero is never conscious of "dying for art," never calls himself an artist, never enounces any flowery ideals. Hamsun could do this because he knew that *this novel itself*—written by the man it was about—would prove that he was an artist.

This correspondence is, essentially, what the film—any film of this book—lacks, since it is not an autobiographical film about a starving young film maker. And it may be that, for this integral reason, the book is really unfilmable—its essence is in its being a book. But to that basic difficulty the adapters, Henning Carlsen and Peter Seeberg, have added defects. They leave things unconcluded: what happened to the revised article that the editor wanted? They omit the walks into the country, which not only reunite the countryman with his origins but emphasize that the bulk of the book consists of his walking, or resting from walking. They omit the verse play that he writes feverishly toward the end. They give no significance whatever to his departure on the ship at the end; it simply happens.

And to all this Carlsen, who directed, adds more defects. Several times he keeps the camera on other characters watching the hero walk away, thus cracking the work's essential solipsism. He has cast as the girl Gunnel Lindblom, an actress whose hallmark is biting passion. When she takes the hero to her apartment and allows him to begin making love, we believe her; when she asks him to stop, we do not. The role needed a more reticent actress, so fascinated by the man as to breach her conservative behavior. Carlsen has some sympathy with the book and has tried, pretty conventionally, to reproduce some of its qualities, but much of the time his camera remains an observer, trying to make an art film. *Hunger* cries out for Bergman. How I wish he had made this film instead of *Hour of the Wolf*. The materials are not identical, but they suggest one another. Hamsun's epic of a rightly mad, lonely artist was made to order for Bergman's gifts of interior habitation. Which is what Carlsen's film generally lacks.

But Oscarsson! Stiff, gaunt, ragged, proud, secretly gleeful, secretly desperate, he stalks through that gray city like a torch whose burning is invisible to everyone but us. That was the prime necessity for the role, I think—the ability to ignite, to light a central flame that is confidently and carelessly there even when he is kowtowing to editors or begging for shelter or raving to old men in the park. He knows who he is. And this is what the girl senses in the unshaved, bespectacled young man with brown teeth and soiled clothes. It is not only sex, though sex he certainly emanates: it is uncomprehended awe on her part. All fully certified by Oscarsson.

In 1966 Oscarsson won the Cannes Festival award for this performance, a singular instance of sense in those prizes. The film was shown in the 1966 New York Festival, then waited two years for

release. To read the Bly translation and then see Oscarsson—for
he has based his performance on the novel, not on the screenplay—is
an unforgettable experience.

The music is by Krzysztof Komeda, Polanski's customary composer, and is very helpful.

Incidentally, I wonder how it feels to be Norwegian and hear this
Norwegian classic spoken in Swedish. Is it like being an Italian and
hearing Manzoni in Spanish?

Warrendale

(September 21, 1968)

Warrendale is so moving, so fascinating and fine, that I hesitate to
say what it's about. The moment I mention the subject, the reader
will perhaps think that the film is noble and worthwhile but that he is
willing to take its worth for granted and spare himself. This would be
self-cheating: not of information or duty but of humanity and, in a
paradoxical way, of joy. Warrendale is a documentary about emotionally disturbed children. It is not a study, it is not propaganda.
It is an experience, passionate and compassionate.

The title is the (former) name of a center in Ontario for disturbed
children, not brain damaged or mentally defective children. In 1966 a
Canadian film maker named Allan King was commissioned by the
Canadian Broadcasting Corporation to make a film about the place.
He spent a month getting acquainted with the children in House Two.
Then he brought in his cameraman, William Brayne, and his sound
man, Russel Heise, for about two weeks of similar visits. Then
they shot film for five weeks in and around the house. Out of forty
hours of footage, this hundred-minute film was edited by Peter
Moseley. Hurrah—just plain simple hurrah—for all of them.

Most feature films are made by men who first create or help create
or somehow acquire fictional scripts, and then guide actors and other
artists to the fulfillment of the fiction. With a film like Warrendale,
nothing can be created except—a huge exception—the confidence of
the subjects. The film makers have to know, really, who their subjects

are, and the subjects have to believe it. In short, the prime require-
ment is not film talent as such, though these men have enough, but
empathy, communion, credibility. The most brilliant film maker alive
would have been powerless to make *Warrendale* without the confidence of those children (and the adult staff). That confidence, in
King and his colleagues, shines from the screen—principally by virtue
of the film's very existence.

It starts with the counselor of the house, a young woman named
Terry, waking the children one morning and having a tussle with a
teen-age girl who refuses to get up, who pulls the blankets over her
head and fights Terry. My reaction the first time I saw this film (I've
seen it twice) was that the girl was perfectly right: who *would* want
to get up when there was a camera grinding away in the bedroom?
And I began to warm up all my prejudices against the intrusiveness
of much *cinéma vérité*. But it didn't take long to see that my feeling
was quite misplaced, that the girl's reaction was (one might say)
natural—she didn't want to get up just as naturally as if she and Terry
had been alone. This is proved by the spontaneity of all the other
actions in the picture, including many by that girl. The camera quite
obviously became just another occupant of the house. At one point,
one of the boys, blithely playing Red Light with some other children
in the street outside, confides to the camera that he can see his
friends' steps with his back turned because of the reflections in the
lens.

The basic Warrendale technique is "holding": when a child has an
emotional seizure, an outsize tantrum, one of the attendants—sometimes two or three—pins his arms and legs and lets him rip. Complete freedom of feeling is the essence, with restraint to keep the
child from hurting himself and to provide a sense of physical contact,
the *caring* of somebody else. We see this method used frequently with
these volatile children. But, crucially, a foreword tells us that this is *not
a documentary about a technique,* it is a personal, selective record of
an experience. I have no idea whether the "holding" technique is good
or bad therapy. I do know that King's film about the place where it
was used brought me close, in a naked and tribal way, to five or six
emotionally disturbed children. It revealed not only the personalities
but the worth of these children. There's a boy named Tony, about ten,
splay-toothed and curly-haired, whose every second expression is
"Fuck off," repeated in a pathetic defensive litany. When he's struggl-

ing in the counselors' arms during one of his tantrums, swearing furiously, I could only think, because of what I knew about him, even because of what he was doing at the moment and why he was doing it, "That's a *wonderful* kid. That's a terrific human being." King had led that boy on to film before then, had shown him playing and blushing and teasing and talking; now, because Tony was *present,* his tantrum seemed one of his ways to express an exceptional sensitivity.

The film merely presents some events in the life of the house. The central point is the sudden death of the relatively young Negro cook, a woman evidently dear to everyone. The chief counselor decides to announce it to all the children at once, and the resulting scene is heartbreaking—but not in a bedlam horror sense. Before the meeting one of the counselors asks the chief how they can explain the death to the kids when they don't understand it themselves. What we see with the children is this bafflement and fright *in extremis.* All the children feel various kinds of guilt for the cook's death. This—enormously amplified in them—is something that all of us feel at sudden death, particularly of the young: not directly responsible, as the children feel, but haunted by the sense that we ought to have been able to do *something.*

This experience is a model of the whole film. These children act out, in exaggerated and baroque ways, many feelings that other children—other people—feel and suppress or understand objectively and can control. These children have little objectivity or control and they just let go: guilt about having been unloved in their homes, as if they had earned neglect, as if they were undeserving of this place and its care; fear to love because of the fear of loss of the beloved; unbridled anger at the teeming mysteries of just one ordinary modern day's existence. Society has not (or not yet) given them the means to control their fears and to invent answers, as it has given to many adults and to the clergyman who presides over the cook's funeral.

Any film that is an impromptu record is likely to have roughnesses and omissions. For instance, it's clear that King was caught slightly short because the cook died early in the filming and he had only a little footage of her. (Understandably, he shifts her death to a point near the end of the film; strict chronology in this matter was not important, and the film would have run downhill if he had followed it.) Some of the sound could be clearer, some of the sequences fuller. A few of the children are left virtually unnoticed, like a pretty teen-age girl, flirtatiously dressed, who sits in the background chewing gum and reading magazines while other children are threshing about in counselors' arms.

But much more bothersome are two extrinsic facts. The first is that the Canadian Broadcasting Corporation, having commissioned this film, refused to show it because it contains—often—the words "fuck" and "bullshit." I hope that at least some members of the CBC felt that this decision was a fucking disgrace. Would it have been impossible to show this utterly humane, basically ennobling film late at night—even if it meant canceling for one evening some acid-in-the-face private-eye thriller with scrubbed language?

Second is the fact that Warrendale has now changed hands and methods, largely (I'm told) because of controversy over this film. I'm as incompetent to comment on the political questions as the therapeutic. I do know that, watching this film and knowing that at least some of the children have been moved and are being treated differently, I felt that something alive and organic and nourishing had been hurt.

Last year we saw a documentary called *Titicut Follies*, made in a Massachusetts institution for the criminally insane, a picture that no doubt originated in a genuine impulse to expose oppressive conditions but that, I thought, began to get some gawking kicks out of showing them. I mention that picture only to assure those who saw it—or who wouldn't see it—that *Warrendale* has not the slightest resemblance to it. It is not an exposé, it is not a chamber of horrors. It is a union with some children who become very precious to us before the hundred minutes are up. Partly this is because they are in themselves interesting and they are allowed—induced—to be *there;* partly it's because they seem to be us, under a distorting magnifying glass. Jean Renoir has called Allan King "a great artist"—not a bad compliment from a man who is a pretty fair artist himself. Inarguably, King has evoked those children's inner selves so powerfully on the screen that he has snared us up there, too.

The Charge of the Light Brigade

(*October 12, 1968*)

THE light on Vanessa Redgrave's cheeks, John Gielgud as Lord Raglan clucking over the British Army like a father hen, the sunny-misty air above smooth English lawns, Trevor Howard as Lord Cardigan turning pink at offense, Jill Bennett as Mrs. Duberly saying, "Oh, glad!" when she and her husband are invited to dine on milord's yacht—these are only some of the things I expect never to forget about *The Charge of the Light Brigade*. It's an extraordinary film, easily the best spectacle since *Lawrence of Arabia* and more am-bitious than Lean's picture since it wants to dramatize not one man's life but a civilization.

Cecil Woodham-Smith's celebrated book *The Reason Why* is given as the source of additional material, but it seems to me the source of virtually all the facts. Even some of the fictional material is adapted or arranged from facts that the book cites. The point of the film is to recreate mid-Victorian England in spirit and detail, to show how that spirit insisted on having a war, and how the most famous battle in that war, by its valor and stupidity, summed up the quality of the age. More than half the action takes place in England before the troops sail for the Crimea; Balaklava, says the film, is what England was heading for, the tea parties and balls as well as the drills and parades.

The picture's absolute triumph is the level of the acting. Gielgud is supreme: his Raglan goes far deeper than the caricature it might have been to become a genteel, gently shrewd, and finally failed man. Miss Redgrave, as a young officer's bride, continues to be magical; she can make us forget that she is acting *at the same time* that we admire her power to make us forget it. Trevor Howard's Cardigan is his best performance since *Sons and Lovers,* a perfect portrayal of a man with the power to exaggerate social conventions into dangerous lunacy. David Hemmings, despite his small size and unexceptional features, has fire and dash as Nolan, the officer who loves his friend's wife. Mark Burns has the right Victorian sweetness as his brother officer. Miss Bennett, as a gamesome officer's wife who follows her husband to war, spices the screen with elegant, nutty sexiness. In all the large cast there is not one false note, but I want particularly to mention Roy Pattison, a veteran sergeant major who is wrecked by believing what the army taught him.

Onward the compliments must roll. The script, by Charles Wood, in its scene-by-scene construction and its dialogue, is intelligent and literate. The sound track, handled by Simon Kaye, is so crisp we can almost taste it. The costumes by John Mollo and David Walker jump from the engraved page into daily use. The color camera of David Watkin is used exquisitely, particularly to capture the soft shine of Victorian photography (as on Miss Redgrave's face). And much praise must go to Tony Richardson, who directed.

I've taken a great many cracks at Richardson since he entered films and I have some more to take here, but a lot of his work in *Light Brigade* is fine. Apparently he has been to school to Lean and *Lawrence,* but that was a good school for this film and Richardson has profited. He has learned especially how Lean plays off huge spectacle against huge intimacy to create a world. I don't mean the easy juxtaposition of vast panorama and tight close-up; I mean the delicacies of individuality poised against the immensity of history.

Richardson has succeeded better with the intimate than the panoramic. For instance, there is a lovely scene in the doorway of a country house where Miss Redgrave stops her lover, Hemmings, who is about to join her husband in the garden, to tell him that she is pregnant. We follow the pair as they step behind the open door, then they move to the front of the door and we see them through the gauze curtain, then we come around to join them, and we remain with her as he finally goes out into the garden. That little sequence of actor and camera movements, not quite parallel, curved in soft arcs, is a mute comment on the poignant helplessness of the pair. When Hemmings lay dead on the battlefield, I remembered his face through the curtained cottage door.

The direction of the Crimea sequences (actually shot in Turkey) is less secure. Richardson here seems more to know what he wants than to be able to get it. But some of this is because large-scale battle scenes are doomed to remoteness. Whether it's Borodino in *War and Peace* or Balaklava here, they always seem to reduce to the same shots in differing uniforms: the *Alexander Nevsky* shot, in which the camera rolls along looking down a line of advancing riders; the cannon exploding in our faces; the quick glimpses of men with lances through their guts; the riderless horses; the ground-level shots of the dead. The big film battle has become a ritual, rather than an experience, often confusing and usually too long. (Having gone to all that expense, they're not going to use only a couple of minutes' footage out of it.) About all that ever really works is the long, wide horizon shot, which conveys only size, not heat. Richardson adds to the difficulty of making the battle *our* battle by including some shots from the Russian viewpoint. They only reminded me that Ensign Tolstoy was involved in this war, and so confused my sympathies.

And this brings me to other shortcomings. Richardson's Social Significance touch has not become exactly light. His contrasts of London poverty and aristocratic *luxe* are plunked down more heavily than they were in *Tom Jones,* the difference between soldiers' squalor and officers' privilege is troweled on. He has also hit on the really abominable idea of using animated cartoon sequences to convey facts and cover transitions. These cartoons blot the visual texture of the film; they distract us and jolt us back to a movie theater; and—worst of all—their blatant satire weakens the self-satire that the film otherwise achieves through realistic fidelity. Who needs cartoons about Victorian stuffiness when we have seen a scandal in an officers' mess because a captain allows wine to be put on the table without decanting? Further, the cartoons oversimplify the causes of the war. British jingoism and wealth and bumptiousness were largely responsible, but another important reason was Liberal revulsion at Russia. It was vividly remembered in England that the czar had helped to suppress the revolts of 1848 and had helped Austria to suppress Kossuth's revolt in Hungary in 1849. This Liberal motive in the Crimean War is left out of the cartoons—out of the entire film.

And finally, *Light Brigade* falters in the intent mentioned earlier— to show that Victorian society itself produced the tragedy at Balaklava. This idea just doesn't fit the facts. The costly blunder was not the result of class privilege or purchased commissions or inhuman treatment of troops. (Indeed we see the survivors of the charge asking to "go again.") It was due, first, to a failure by the goodhearted and thoughtful Raglan to make himself clear and, second, to an ambiguous statement in the heat of battle by Captain Nolan—the most progressive and humane officer of the lot! The Crimean War, after Britain muddled through, did eventually bring about various reforms, but the fiasco of Balaklava was not directly caused by the ills that were later reformed. So, thematically, the picture is inconclusive and leaves us feeling itchy, unsatisfied.

But if it's less than the sum of its parts, many of those parts are splendid. It's not to be missed and, I think, not to be forgotten.

Weekend

(*October 19, 1968*)

By far the best pro-Godard commentary I know is Susan Sontag's essay in the Spring 1968 *Partisan Review*. I reread it between my two viewings of *Weekend*. Miss Sontag sketches the adverse criticism of Godard ("What his detractors don't grasp, of course, is that Godard doesn't want to do what they reproach him for not doing"); then she examines these points, attempting to show that the supposed faults are part of Godard's method. Miss Sontag overlooks the fact that some adverse critics assume that Godard works with intention but that intentionality does not itself create an esthetics; still her arguments, all relative to *Weekend,* are the best critical support for Godard that I can imagine.

On the matter of Godard's flashy use of ideas and of literary references:

Certainly ideas are not developed in Godard's films systematically. . . . They aren't meant to be. In contrast to their role in Brechtian theater, ideas are chiefly formal elements in Godard's films, units of sensory and emotional stimulation. . . . What's required is that literature indeed undergo its transformation into material, like anything else.

This does not exactly contravene the objection—by many, including me—that Godard is irresponsible in his use of explosive political ideas and callow in his literary display. It says that he is masticating these matters into fodder for cinema; that he treats, say, Mao and Dostoevsky as he would treat a tree, a flower, a kiss. I think this approach is antihistorical, antiintellectual, and finally anticultural, but it does have an imperial bravado.

On the ceaseless display of Godardian "effects":

. . . Godard proposes a new conception of point of view, by staking out the possibility of making films in the first person. By this, I don't mean simply that his films are subjective or personal. . . . [He] has built up a narrative presence, that of the film-maker, who is the central *structural* element in the cinematic narrative. This first-person film-maker isn't an actual character in the film. . . . He is the person responsible for the film who yet stands outside it as a mind beset by more complex, fluctuating concerns than any single film can represent or incarnate. . . . What he seeks is to conflate the traditional polarities of spontaneous mobile thinking and finished work, of the casual jotting and the fully premeditated statement.

That is a sympathetic description of Godard's effort to make every film a record of his experience in making the film, of the tension he wants to convey between the film and the world, of his frenzied insistent drive to treat film as if it were not a photographic record, fixed before we see it, but something happening at the moment we see it— a response to everything around the film and in Godard at every moment. (In *Weekend* there is a brief flash of some Italian actors "stranded" from an Italian-French coproduction. *Weekend* is such a coproduction. The shot is a spontaneous impulse, irrelevant to the story, an attempt to get the "outside" of the film inside.) What Miss Sontag disregards is that even the *Divine Comedy* was created by a mind beset by more complex, fluctuating concerns than that poem could incarnate, that Godard's struggle for seeming spontaneity is doomed because no film is a spontaneous event and because the effort to seem spontaneous can get wearisome. With Godard we become aware of the desperation, of the *fixed and photographed* impromptu.

I cannot summarize all of Miss Sontag's article (it should be read), but, for me, it leads to and away from this sentence:

Just as no absolute, immanent standards can be discovered for determining the composition, duration and place of a shot, there can be no truly sound reason for excluding anything from a film.

This seemingly staggering statement is only the extreme extension of a thesis that any enlightened person would support: there are no absolutes in art. The Godardians take this to mean (like Ivan Karamazov) that therefore everything is permissible. Others of us take it to mean that therefore standards have to be empirically searched out and continually readjusted, to distinguish art from autism; that, just as responsive morals have to be found without a divine authority if humanity is to survive, so responsive esthetics have to be found without canonical standards if art is to survive. The last may be an open question, but it *is* open as long as men continue to make art.

Godard is grateful, I hope, to Miss Sontag for her effort to give esthetic order to a method that is essentially impulsive and whimsical. Her arguments are ingenious and erudite, but I think they end up as rationalizations, not analyses. Take *Weekend,* for example. Miss Sontag would say that it purposefully "conflates" casual jottings and premeditated statements. To me it is a film by a prodigiously gifted man who had a general plan, held it to for a while (with some stunning results), then became impatient and increasingly self-indulgent to keep from being bored.

Weekend begins as a relatively formal social satire, very sharp. A Parisian couple are going to drive to the country on the weekend to see the wife's rich mother and try to get a share of her estate. They are venal, callous, disloyal. Each secretly loathes the other, each has a lover. (In an early scene the wife, in bra and panties, describes at dreamy length a sex orgy she had with another pair—while her lover takes notes. This may be a takeoff on Bibi Andersson's sex story in *Persona;* if so, it proves that Bergman understands sex better.) Before the couple start out, there are some minor car accidents with disproportionately furious quarrels, very funny. They run into a huge traffic jam in the country, again very funny, and this sequence ends with a reversal of the opening: a dreadful accident that is taken lightly. Cars matter; people don't. More and more bad accidents appear on the edges of the film and begin to sift into the center, until the couple are adrift in a world gone mad with automotive homicide.

Through this point there is plenty of Godard apparatus: visual puns, the characters say they're in a movie, lots of literary references (including Joseph Balsamo out of Dumas), and so on. But it's a consistently developing story. Only about midway does the film really become "first person" in Miss Sontag's sense as it begins to include "everything." The turning point is a scene in which a pianist plays a Mozart sonata in a barnyard (no, I'm not going to explain that) while the camera executes two slow 360-degree pans that have already made some movie buffs gasp with admiration. From there on, the film sweeps up everything that Godard encounters or imagines as he tells something of a story about the disintegration of society because of the breakdown of cars. We see the rise of young revolutionaries dressed as Indians who kill and die with the aplomb familiar from *Les Carabiniers*. Politics gets spouted, by an Arab and a black, but these speeches, Miss Sontag has warned us, are not to be taken seriously as politics—they are merely part of the film's materials, included because they are in the world.

Weekend goes on too long, which means only that Godard's improvisations and spontaneous responses on his theme grow tired before he stops. The earlier razor-slashing satire has long since disintegrated into a grabbing at straws of shock. But much of it is Godard's keenest conception and best execution, the present-future of *Alphaville* turned into the future-present.

Romeo and Juliet

(November 2, 1968)

FRANCO Zeffirelli, who directed *Romeo and Juliet,* is temperamentally a Victorian. His idea of Shakespearean production is to heap on *things*—fabrics, scenery, lighting effects, complicated props—in a way that would have delighted Mr. and Mrs. Charles Kean at their London theater in the 1850s. If only the Keans could also have had real *places,* as Zeffirelli has, they would have been deliriously happy.

So Zeffirelli is better off on film than he was with his Old Vic production of *Romeo.* On stage, clever as many of the effects and mob scenes were, the cleverness was quickly outweighed by the inadequate rendering of the play itself. The same thing happens in the film, but more slowly. Here Zeffirelli has chunks of Italy to interpose, and more extensive acrobatic swordplay and plentiful gorgeous old architecture. All this not only postpones our discovery of the hollowness of the production, it gives him a valid excuse. "Well," he says figuratively, "you can't have everything. Look at how much Renaissance I gave you. The two hours are almost up. Have to cut the text a bit, if you don't mind." Many Victorian productions whittled the play, too, though not in the same way. They made cuts to magnify the stars; here the cutting is done to give the stars less—very little of the "Bid me leap" speech, for instance, and not one word of the potion scene. Also, no apothecary scene, no Paris in the tomb, etcetera. The text is fairly well held through the midpoint, then we are whooped along to a conclusion.

As a chef of feasts for the eye, Zeffirelli is Cordon Bleu. (Much better here than in his previous abomination, the Burton-Taylor *Taming of the Shrew*.) As a director of Shakespeare, he simply doesn't exist. He's not an innovator like Peter Brook, he's an evader. He has no ear, no insight, no *interest* in the delicate centers of the play. He supplies a lush pageant that uses Shakespeare's story and large patches of the text. But in terms of a drama that is intricate with ideas of fate, a tragedy that is contained in its language, not in a synopsis of its plot, this just is not *Romeo and Juliet*.

In a sense this picture only proves again that the film medium and Shakespeare are born antagonists; no Shakespearean film has ever proved otherwise, though some have worked out better truces than this one. But Zeffirelli is not just a hapless victim of esthetic circumstances. He is a directorial charlatan who thinks that, by a lot of Tarzan tree climbing in the balcony scene or by Mercutio's splashing around in a fountain, he is revitalizing a poor old musty play. Certainly there are other ways to do Shakespeare besides the traditional, but *not* to do him isn't one of them.

The crux is in his casting of the two principals. Zeffirelli selected two unknowns, Leonard Whiting, seventeen, and Olivia Hussey, sixteen, who are billed as "the youngest performers ever to play the roles professionally." Nonsense. For just one example, Master Betty played Romeo at Covent Garden in 1805 when he was thirteen (see Giles Playfair's *The Prodigy*). His purpose was to get appropriate youth in those parts, which indeed is usually lacking. He succeeded; in fact, some of the few poetic moments in the film are close-ups of Miss Hussey's face—magnificent eyes—as long as she keeps her mouth shut. Unfortunately this young pair often open their mouths. They have tinny voices, harsh accents, don't understand much of what they are saying, they haven't a clue about reading verse, and they are abysmally amateur actors. They epitomize Zeffirelli's approach to direction: gorgeous surfaces.

Milo O'Shea bumbles along as Friar Laurence, very little different

from his Bloom in *Ulysses*. John McEnery's Mercutio is a street corner showoff, not the quintessence of courtier's panache. Michael
York and Paul Hardwick are acceptable as Tybalt and Capulet. The
best performance is the most conventional one—Pat Heywood playing the Nurse with the stock cockney accent. Robert Stephens, as the
Prince, struggles nobly under a funny hat.

Funny Girl

(November 9, 1968)

MILLIONS of people will now see Barbra Streisand whole. Thousands
have seen the whole Streisand on Broadway in *Funny Girl*, but many
more have seen her only in TV specials that sliced her neatly down
the middle and presented only the singer, the fashion plate and dish.
In the film of *Funny Girl* they will also get Streisand the comic—the
Jewish comic, a superior Jackie Mason in drag—and will thus understand her immense success, which may have been puzzling them.
As a singer she is good, but not good enough to keep from being a
bit tedious in a long special. Her singing, which is mostly à la Garland,
seems much better as the additional talent of a good comic, and her
comedy seems better when we know she is a good torchy singer. In
one person, Punch and Judy.

The film itself is rotten, but so was the Broadway show. The songs
are mediocre, the book worse. (It's based, need I say, on the life of
Fanny Brice.) William Wyler, a generally able director, has done his
creakiest work here, full of slow sentimental dissolves and strained
attempts to make like Minnelli and Donen (for instance, different
stanzas of a song sung in different locales). As for the "straight"
sequences, it's hard to believe that the man who controlled the hokum
so beautifully in the last scene of *Roman Holiday* could have been so
paralyzed by hokum here. Harry Stradling, whose camera ladled
Technicolor gravy all over *My Fair Lady,* keeps ladling along. Omar
Sharif plays Nicky Arnstein. What a performance! Isn't there enough
trouble between Jews and Egyptians?

The whole feeble enterprise needs to be *carried,* and the point is that Miss Streisand is well able to carry it. She's a bit irritating because she is evidently conceited—evidence that filters through from the screen, not from the press—and it's always irritating when a conceited person is as good as he (she) thinks. A further point: the social importance of Miss Streisand's face. Now we're being told—with a predictable reverse twist—that she is not really homely, that her talents make her beautiful. This is just word juggling. What is the theme of the show other than a homely girl's problems and triumphs? Why else was Miss Streisand, a relative unknown at the time, cast in the Broadway show originally? And she is *Jewish* homely. To disregard both these elements is to disregard the importance of Miss Streisand's emergence, not only as a star but as a sex figure. This would have been impossible forty years ago for Fanny Brice herself, who was at least as talented as Miss Streisand and rather prettier.

Beyond the Law

(November 16, 1968)

I DON'T really want to review Norman Mailer's new film, but I'd like to say why. *Beyond the Law* is one more piece of almost maniacal self-indulgence, abetted by Mailer's idolatrous friends, without merit or interest except as it affords a peek, for those who care, into the private games of Mailer and pals pretending to be cops and criminals and hippies. (George Plimpton appears briefly as himself pretending to be the Mayor who visits the police station where a lot of it happens. Alan Alda is better as Plimpton, in *Paper Lion,* than Plimpton is.)

Those who are dedicated to finding value in everything that Mailer does can find more of it here, I'm sure. Those who find something of *vérité* in every foot of film that's exposed by egotistical amateurs can write their reams about this picture's relevance to role playing in modern life. Good luck to them. I take this film simply as part of the price for having Mailer around. He splashes about, he jostles, he elbows, he irritates; in his ambition (announced at the last New York Film Festival) to be a Renaissance man, he overlooks one small detail —the archetypal Renaissance man, Leonardo, didn't merely work in several fields, he was good in all of them.

But Mailer is also the man who wrote *The Armies of the Night* and *Miami and the Siege of Chicago* in one year. Those books—particularly the first—are not exactly free of arrogance and solecism, but they are the outpourings of a potentially large writer who has at last found his moment. Ever since the Second War, about which he wrote the best American novel, he has always been just out of step: a Depression child when the Depression mind-set no longer applied; a Jew when it became difficult to suffer in America for being Jewish; a white man when the drama became black; a middle-aged man when the action became young. With the Pentagon march last year and the political conventions this year, the man and the moment coincided, and as an author, he became what he has always longed to be, an agonist-prophet of our time, embodying many of us in his ego, possessed by burning insight.

It's silly to wish he would stop making films, as it would be silly to wish he would stop writing his time-filling fiction or embarrassing articles like the defense of his friend Podhoretz's *Making It* in *Partisan Review*. One might as well wish that he would stop getting drunk in public. His films are part of the same urge, I think—a frenzy to grab at everything, every possibility of sensibility and power: which he tries to hide under the label "Renaissance man." I don't review his adolescent binges in life, and I don't see any reason to review his Dreams of Glory films. But after I saw *Beyond the Law* and was ready once again to write off Mailer as one more man exploded out of his head by the sheer size of contemporary experience, I read the Miami-Chicago book. Here's a passage in which he explains why Chicago people remind him of the people in Brooklyn where he grew up:

. . . they were simple, strong, warm-spirited, sly, rough, compassionate, jostling, tricky and extraordinarily good-natured because they had sex in their pockets, muscles on their backs, hot eats around the corner, neighborhoods which dripped with the sauce of local legend, and real city architecture, brownstones with different windows on every floor, vistas for miles of red-brick and two-family wood-frame houses with balconies and porches, runty stunted trees rich as farmland in their promise of tenderness the first city evenings of spring, streets where kids played stick-ball and roller-hockey, lots of smoke and iron twilight. The clangor of the late nineteenth century, the very hope of greed, was in those streets. London one hundred years ago could not have looked much better.

OK. If we have to have Mailer's films (he has just finished another) as part of the price of an author who can cascade like that, OK.

Yellow Submarine

(November 16, 1968)

LOVELY Beatles. They not only do beautiful things themselves, they have a gift for finding other gifted people. When they went into films, they found Richard Lester, exactly the right director for them. Now they have produced an animated cartoon feature, and they have found a staff of artists under Heinz Edelmann to do the drawings and collages. I don't know how I knew what a Beatles animated film ought to look like, but the moment *Yellow Submarine* began, I knew Edelmann and Company were exactly right.

The screenplay is by four authors. This must be because four different fancies had to dream up the assortment of odd creatures that tumble through. It couldn't have taken four men to devise the plot itself, which might have been invented—and often has been—by your little nephew Willie. The story opens in idyllic Pepperland, as it is invaded by Blue Meanies. The conductor of the Sergeant Pepper Band flees in a Yellow Submarine, which surfaces in Liverpool and takes the Beatles aboard. There are wild encounters with various wild characters: Snapping Turks, whose waistbands yawn wide to disclose rows of shark teeth, the Ferocious Flying Glove, from whose index finger come lightning bolts; the Count Down Clown with the suction nose, and many more. There are lots of well-known Beatles songs on the sound track, including the two that I guess are my favorites, the title song and *Eleanor Rigby,* and there are three new numbers.

Fine. The only discomfort in the film is a slight disproportion in the Beatles' basic mixture. What has always charmed me about them is their combination of blitheness and compassion. Their cool about loneliness and stupidity and cruelty makes their songs on these subjects all the more effective. Here some of the story and the trappings—particularly toward the end—get softened with a dash of

Disney. Love probably *is* the best thing that our culture has invented, but the Beatles have usually managed to say it airily. Here the message sometimes plops a bit.

Well, let it plop. This film, like so much that the Beattles do, is both unexpected in their careers and delightful. If you have kids, prepare to share them now. *Yellow Submarine* really fulfills the oft-stated claim—it's for people of all ages. To be able to make that kind of picture must take some rare quality of soul.

Bullitt

(November 23, 1968)

MAYBE the police film is trying to tell us something. In the twenties and thirties such a film was often about an immigrant family, usually Irish, with a father on the force and at least two sons. One of them followed in Pop's beat-steps and became a good cop; the other became a cop and went crooked, or else saved time and went crooked at once. In either case the last scene was brother against brother, and you know who won. Overtly the story told us about the virtue of virtue, and, underneath, it was thrumming away about the sure rewards of conforming to the American Dream.

The police film still tells us about keeping noses clean and badges bright, but with a more complex basic message. Even before we get to basics, there are lots of surface changes. The police commissioner in *Madigan* has a mistress, and Richard Widmark, in the title role, has new-fashioned marital troubles. Frank Sinatra in *The Detective* has a nympho wife whom he loves hopelessly. Jewish cops, with Jewish mamas and wives, are very much in—George Segal in *No Way to Treat a Lady* and Jack Klugman in *The Detective*. Presumably this is to show us that Jews are neither cowards nor job snobs. Lots of other details and plot twists—race conflicts and candor about homosexuality, for instance—help to bring these films up to date.

But these are easy accommodations, in milieu and minutiae. The *fundamental* view has also been changing, and this is neatly exemplified in *Bullitt*. Put simply, the new view is the policeman as private

eye. Steve McQueen plays Bullitt, a regular detective of the San
Francisco police, but every atom of his being and a great deal of his
behavior belong to the loner, the cool Hammett-Chandler-Bogart cat.
(They all collaborated on the persona.) The private eye was the
realist's hope for intelligence and some justice against the uniformed
forces of stupidity. He stood in relation to the police as the man who
really sees the score against those who see only the rulebook, and his
life style suited his life view. Now we have McQueen, a plainclothes
cop, dating a swinging chick with whom he visibly beds, driving a
sports car (borrowed), living in a pad, dressing sharp, and, like
Sinatra's detective, making his own rules as far as possible. He looks
and lives like an independent operator.

One drift of this change may be to make respectability respectable,
to show (particularly to the young) that law and order are not
necessarily Dullsville. Underneath this possible propagandistic motive,
however, the change dramatizes another, deeper tide of the times: not
so much the liberalization of the square as the engulfment of the free.
Just as the most avant-garde art these days is quickly swamped with
suburban acceptance, so the cynical private eye—fighting for truth,
sometimes against the nominal forces of truth—gets engulfed by the
civil service. The last line of the official Warner Bros. synopsis: "By
good detective work, Bullitt has cleared up the case while maintaining his integrity in the face of a threat to his career." No PR man
could ever have written that about a Bogart film. When Bogey cleared
up a case, it was so he could be free to tell anybody to go to hell. The
real finish of the *Bullitt* synopsis is: Be reasonable. See how enlightened we are these days. You can have freedom—a reasonable amount,
anyway—inside the organization. Be reasonable.

In the Broadway-Hollywood phrase, the picture is "smartly
produced." It uses modish décor and much modish film technique,
with about equal seriousness. The script is odd in one respect. It's
full of sharp TV backchat, but its central gimmick is obscure, and the
authors make the hero professionally stupid. This fellow who lives
like a panther doesn't have the police brains of a backwoods rookie.
When he is warned that a killer is loose in the hospital where he is
guarding a wounded man, he bumbles the protection. When he knows
that a dangerous fugitive is on a plane at an airport, he doesn't call
for men to surround the plane. And there are further phoninesses:
social-political comment is laid on with high-school irony in the per-

son of an ambitious politician. (At the end this man rides off in a limousine and unfolds the *Wall Street Journal!* Just the sort of paper a rat like that *would* read.) And there is a long high-speed auto chase through San Francisco, full of visceral quivers but quite false because, as I remember, not one pedestrian was visible, even on the sidewalks. Such a chase through the normal streets of San Francisco would have ended in deaths much sooner than it does.

The director, Peter Yates, is clever enough to give the whole film a smear of style, none of it his own. McQueen moves lithely through his part, depending quite a lot on the flatness of his stomach.

Faces

(December 14, 1968)

JOHN Cassavetes has made two independent films—*Shadows* (1960) and now *Faces*—which are concerned with unvarnished truth, so it's curious that in both of them he does some varnishing. *Shadows* maintained that it was entirely improvised, but this seemed questionable. *Faces* is very much better; still there are things in it that contradict its air of candor and brute fact.

Cassavetes (who is also an actor—currently to be seen as Rosemary's husband in *Rosemary's Baby*) spent several years making *Faces* in and around his Los Angeles home. The theme is well-heeled America's frustrations, sexual and spiritual, their sublimations and palliatives. This is John O'Hara country, and Cassavetes' method is a film cognate of O'Hara's process of selective eavesdropping.

The nub of the story is one night spent apart by a middle-aged couple who have quarreled. He sleeps with an expensive call girl. The wife, with some matronly friends, picks up a stud at a go-go joint, and the stud chooses to stay with her. In the morning the husband comes home, the stud (a rather goodhearted fellow) flees, and the couple are once again under the same roof. But only literally. Cassavetes has used a hand-held camera most of the time, lighting that deliberately looks like hurried TV-newsreel setups, and editing that strives for a sense of snatch-and-grab. He wants to create a feeling of intrusion into actuality, and he often succeeds. The sexy itching restraints of semidrunk businessmen, always on the verge of violence or self-pity around call girls, are microscopically well done, if over-done. The long scene with the stud and the four ladies is obscene, not in what happens but in the degradation most of them feel at what they wish would happen. The silly jokes between the husband and wife convulse the pair with their very silliness, a currency of inti-macy between two people who know each other well. Almost all of these scenes go on too long, but a good deal is admirably unsparing, disgusting, poignant.

Yet for all the *cinéma vérité* thrust of the film, the best element in it is not factualism but contrived art: the performances of the couple by John Marley and Lynn Carlin. Not all the hand-held cameras or jagged editing in the world could provide the reality *created* by Marley and Miss Carlin. (She has never acted before; I certainly hope she acts again.) And the difference between texture-truth and perform-ance-truth is shown in reverse by Gena Rowlands (Mrs. Cassavetes) as the call girl. Miss Rowlands is photographed with the same *vérité* but comes out embarrassed and unconvincing. (Not the character's insecurity, the actress's.)

Aside from Miss Rowlands and the scenes that Cassavetes lets run too long, his ruthless honesty seems to me somewhat compro-mised. There is a prologue in which the Marley character goes to a screening room to see a film—he's in the picture business—then sees *Faces,* with himself, the character, in the leading role. There is no return or conclusion to this idea. This worlds-within-worlds touch seems heavy symbolism in a naturalistic film. The quarrel between the married couple that drives them apart for the night seems trumped up and thin, insufficient to produce such drastic results in this couple. And those results—each one finding a mate at once—have the in-stant symmetry of any factitious script. The call girl has that inter-changeable heart of gold that is transplanted from one fictional whore to another with an ease that must make Dr. Christiaan Barnard green.

Still a good deal of *Faces* is highly effective, even valuable as social history. Marley and Miss Carlin supply some of that history in their performances, Cassavetes supplies some with his seemingly snooping camera.

Up Tight

(December 21, 1968)

JULES Dassin's *Up Tight*, financed by Paramount, is the first big-money film (as against several underground films) to tell uncompromised truth about black feelings in this country today. Not all the truth (what is all?); but, as far as I know, every aspect of black thought that this film treats is treated candidly. It's quite an experience to hear the political sentiments in *Up Tight* coming from a film that begins with that same old Paramount insignia, the lofty peak circled by calm stars.

The picture takes place in Cleveland and begins on the day of Dr. King's funeral in April, 1968. The camera pushes into the black community of the city—very few whites in the film—probing the varied reactions. (There is no mention that Cleveland has a black mayor, but for those who remember this fact, the implication is that, for militants, Mayor Stokes is irrelevant.) The militants feel that non-violence has brought about its own demise in violence, and they are preparing an armed outburst. A veteran black "gradualist" tries to reason with them. A white radical who has marched in Alabama and been jailed in Mississippi is expelled simply because he is white. (Later on, we get a glimpse of the come-to-Jesus religious escape hatch as the script tries to touch every element of black feeling.) The film moves unequivocally toward a finale of revolutionary resolve, with inevitable race war as its conclusion. Radical social revision would obviate that conclusion, but the militants don't expect that from whites. *Up Tight* tells it as the militants feel it must go. There is no Stanley Kramer on hand—or, for that matter, no Bill Cosby or Sidney Poitier—to make it all come out merely ironic or possibly rosy. The heated irons are there.

But *Up Tight* is based on Liam O'Flaherty's novel *The Informer,* from which John Ford's famous 1935 film was made. I say "but" because I think that this is the new picture's undoing. Dassin and his collaborators on the script, Ruby Dee and Julian Mayfield, have felt bound to follow the O'Flaherty-Ford form; thus what starts as a dramatic conspectus of present tensions shifts into a drama of one man that is basically unrelated to those tensions. The three new writers have overlooked the fact that Ford's film was not about the rights and wrongs of the Irish Revolution or the conflicting ideological strands within it. It was concerned entirely with its protagonist, who, as far as his drama was concerned, could have been an Algerian under the French or a Palestinian Jew under the British. The O'Flaherty-Ford theme was the humanity of weakness and the futility of betrayal, not social-political issues. *Up Tight* begins with issues; then, in unfortunate obedience to the earlier structure, swerves right out of them, into another subject. Dassin's protagonist, Tank Williams, the renegade militant, does not have a race problem essentially: he has a drink-envy-weakness problem which could have landed him in an equivalent mess if he had been a white defense-plant worker with Soviet agents nibbling at him. Near the end, the hunted Tank says he wishes that someone in the world could tell him why he committed his treachery. This plea is made the crux of the film, but Tank's internal anguish is a long way from the wide-scale social drama with which the story began. The idea of a remake of *The Informer* is presumably what launched this project and made it possible: presumably a much more exploitable idea than a new script on the race situation. But a new script would not—necessarily—have forced the film into a shape that works against it, as this remake does.

What is even worse, this protagonist and his drama are feeble. We are not only switched out of what is valid and vivid; what we are switched *into* is poor. In the second half of the film, the screenplay becomes mostly a weak Americanized echo of Dudley Nichols' *Informer* script, and the performance of Tank is boring. Julian Mayfield, who plays the role, is a big sweating man with no real power of agony. Victor McLaglen was hardly a sterling actor, but he had a forceful personality and Ford knew how to use it.

Some of the cast are good. Frank Silvera, the gradualist, gives fire and contempt to his dissolving position. Raymond St. Jacques, the militant leader, is utterly convincing, never a trite film bravo. Although Roscoe Lee Browne, as a police informer, is blatant, his role is blatantly written—a blind man could see that he is a stool pigeon; at least Browne plays it with bitter style. The least flavorful performance is by Ruby Dee, pallid as Tank's prostitute girl friend.

The use of Technicolor seems wrong. (Perhaps this was Paramount's fault, looking ahead to future TV sales.) *Up Tight* cries out for black and white, to avoid any possibility of the very trap the production falls into. As designer, Dassin has used Alexandre Trauner, the Frenchman whose career of more than thirty years includes *Children of Paradise* and the recent *A Flea in Her Ear*. (His *art nouveau* sets were the best feature of that last disaster.) When Trauner designed such grim French works as *Quai des Brumes* and *Le Jour Se Lève,* he worked out of and toward something he knew. Here he designs like a fascinated tourist. He overdesigns; his colors are lush and intrusive, many of the places look like settings (the room where Tank is "tried," for instance), and the overall visual effect is of a musical.

Dassin's direction is his best since *He Who Must Die,* but that's a tiny compliment because most of his work since then has been abominable (*Phaedra, Topkapi, 10:30 P.M. Summer*). At least *Up Tight* has allowed him to achieve a bit of the spaciousness of his Kazantzakis film. Some of the better moments—like a scene where tenement dwellers crowd onto balconies and toss bottles to protect a fugitive from the police—are reminiscent of early Fritz Lang. But Dassin's grip is weak. He grabs suddenly and frantically for subjective effects: a hand-held camera when Tank goes into the police station to squeal and a whirling camera when the fugitive plunges to his death, a device that's repeated when Tank dies. Dassin reaches his nadir in a penny arcade where Tank encounters some white people in evening dress. The white "backlash" dialogue is sophomoric, and Dassin shoots his actors with equal subtlety—in fun house distorting mirrors. It's enough to make you doubt that any honest statements *have* been made in the film.

The honesties are there, however. *Up Tight* is, in sum, a poor piece of work, but at least it allays our initial suspicion that it's going to crimp its explosive underlying ideas. Which leads to the next question: Why does a big studio, a subsidiary of a huge industrial combine (Gulf & Western), finance a frankly revolutionary film?

For two interlocking reasons, I think. First, there isn't a militant statement in *Up Tight* that will be news to readers of *Look* and *Life* or to viewers of CBS and NBC and NET. In the past, to tamper with matters like the militants' pronouncements would have made the film look phony only to readers of books and of small-circulation magazines. No longer so. The mass media now reach wide, and they can reach fast. They have biases and limitations, but they do have some sense of the dramatic; and nothing is more dramatic in America today than black militancy. TV coverage in particular—the visible actions, the audible statements, the resolute faces—have made Hollywood equivocation on this subject virtually impossible. If a picture is going to deal at all with a black group that is willing to kill a traitor, that group cannot be shown as working toward, say, the bussing of schoolchildren.

Moreover, revolutionary ideas of all kinds have become a new spectator sport for the increasingly affluent, increasingly educated, therefore increasingly guilt-ridden (and guilt-enjoying) middle class. I was in Vancouver recently, and the first thing I saw on the morning I arrived was a hippie on a street-corner selling the local "underground" newspaper, doing a brisk trade with white-collar types on their way to work. Not in San Francisco or L.A.—in Vancouver, British Columbia. Whatever Dassin's motives for wanting to make *Up Tight,* Paramount's motives were, I would guess, the knowledge that the trails had been well blazed by mass media, an awareness of the new sophistication in ideological thrill-seeking, and a professional sense of the narrowing gap between the front line of social fact and the threshold of film-audience acceptance. Plus the money of the growing black audience. In any event, the fact that Paramount backed this picture is more interesting than the picture itself.

Shame

(January 4, 1969)

INGMAR Bergman has made a film about something he hates, war, with people whom he loves, Max von Sydow, Liv Ullmann, Gunnar Björnstrand, and the cameraman Sven Nykvist. This basic contrast, subtle and pervasive, is stronger than anything else in the film. Of course that contrast will be effective only with those who have some familiarity with Bergman's work. Well, I can't speak as someone who *doesn't* have such familiarity, and by now there cannot be many readers who don't have some acquaintance with Bergman, some sense of the accrued richness in his films through his reuse of persons and places. (*Shame* is set—again—on an island off the Swedish coast.)

To indulge this personalist criticism a bit further, the best bitterness in *Shame* comes from a feeling that Bergman conceived this script at least partly by imagining these actors, who are dear to him, being plunged into physical horrors; imagining the cameraman, who has captured so much delicacy for him, slamming down images of brutishness; and imagining the pearl-gray world of the Swedish islands, which have become Bergman's outer landscape of inner landscapes, ravished by flames and smashing. Those contradictions are the fundamental obscenity.

Bergman's method is bioptic. Instead of an antiwar film on a panoramic scale, he tells the close-up story of one civilian couple. It is 1971; no places or armies are named. On this island live a husband and wife (von Sydow and Miss Ullmann), violinists who have retired here because the war has broken up their orchestra and because he has a tricky heart. At the start they are a couple with wrinkled, well-worn domestic patterns: he is a bit cranky, she is a bit cruel; he has his morning pill, she has her morning domination. But they love each other—periodically, at any rate. The waves of military filth gush and countergush over them, as invaders attack and are repulsed. (The director Vilgot Sjöman is briefly visible as an invader officer, I think.) By the end the couple have reversed roles; he dominates. The man who could not kill a chicken has killed a friend and a young unarmed soldier. The wife accepts these killings; had slept with that friend in order to protect their lives; has retreated for refuge into dreams of the child whom the marriage has not produced. The civilized people of the opening have become little more than survivors, whose chief ethic is to survive.

The three leading performances are perfect. Von Sydow seemingly doesn't know how to make a false move, and he makes the true moves so fluently that the design and skill in his acting are easy to scant. Björnstrand is the friend, the mayor who goes from power to sexual enslavement to execution; he has a moving impassiveness, a disbelief in the reality of any event even while that event is clutching him. As for Miss Ullmann, whose third Bergman film this is, she now stands with Bibi Andersson and Vanessa Redgrave in the front rank of the world's young film actresses. She makes every moment crystalline, the quintessence of what it is about. I was about to praise the scene in which she sees her first dead body, that of a child, but that moment is no better—simply more dramatic—than those in which she frets at her husband or loves him or realizes what is happening to him and her or begins to go morally numb. Her face is snub-nosed, freckled, funny; it is also purely beautiful. Which means of course that *she* is funny and *she* is beautiful and has the talent to shape the character with her funniness and her purity.

Bergman knows how to rely on all this. The opening scene of *Hour of the Wolf* consisted of Miss Ullmann coming out of an island house, sitting at an outdoor table, and speaking directly to us for some minutes, a scene that was a triumph of her ability and his judgment. In *Shame* there is a similar scene at a similar table where she speaks at length to her husband (we watch over his shoulder), progressing from wistful unsatisfied maternity to immediate physical desire. Again she plays the long scene excellently; again Bergman shows us that masterly direction sometimes consists of not intruding, of allowing the right performer and the fact of film to come together.

Bergman, with Nykvist, has devised pictorial beauties that dramatize visually what every scene is about, while the actors provide the correlatives: a close pyramidal shot of the husband and wife holding his old violin as they talk of their past musical lives; the long stretch of shore, like the edge of existence verging on nowhere; a scene by the side of a fast-flowing stream (like one in *Winter Light*), where the rushing water mutes the voices. Bergman insists on making physical textures speak—like the contrast between the weathered wooden siding of the house and Miss Ullmann's fresh face in the doorway. And the scenes of violence were made with a hand-held camera, to punch us with immediate shock.

And yet (an antithesis that recurs in my Bergman reviews) . . . and yet *Shame* disappoints. First, it's too long. It says everything it has to say by the time the husband has killed his friend. (In fact *Shame* is about twenty minutes longer than most Bergman films since *Smiles of a Summer Night*.) Some of its ironies are facile, like a sequence in a shop full of eighteenth-century curios or the children's drawings in a schoolroom being used for tortured prisoners. But even if the picture were pared, it would still have too little to say—and by "say" I certainly don't mean "utter" or "preach." It scrutinizes two people slammed about by war, not just to tell us that killing is horrible but also that living through the killing is horrible. *Shame* accomplishes this; nevertheless it is too monodic, almost didactic, too bare of new evocation. Given these people and what is happening, we know

what to expect. What we want from an art work about war—at this
late date in art works about war—is what we didn't know we expected.
Richard Gilman said in *The New Republic* (November 25, 1967)
that the relevance of Richard Lester's *How I Won the War*

is to the way war, or any phenomenon of violence, proceeds in its
usurpation of truth and consciousness. And it shows this by extending the
film as creator of counterstatements, fecund myths whose weight may be
used against all the destructive ones.

Such myths are fecund, when they are, not so much, perhaps, because an art work creates them as because they were in us, fertile
and waiting, and are now touched to fruition by the artist. Bergman
only tells us—often beautifully—what we are already aware of, in
fact and myth. *Shame* forges no new armor, touches no new core,
blazes no deeper shame.

The Birthday Party

(January 4, 1969)

ROBERT Shaw is first-rate as Stanley in the film of Harold Pinter's
The Birthday Party. (Sometimes, with glasses, curly hair, and nar-
rowed eyes, he even looks like Pinter.) Shaw has screen-filling presence and the power to suggest memories echoing in a long, arched,
troubled mind, together with a hint of violent madness just below
the surface. Dandy Nichols is excellent as Meg, his landlady—affectionate, almost moronically innocent of the way he is using her as a
punching bag and, by virtue of simpleminded devotion, achieving
some stature.

The film is worth seeing for those two performances but not for
much else. The play itself is Pinter's first full-length work (1958) and
seems to me his least good long piece. The agencies of threat are too
explicit; the dramatic form is basically a gangster comedy-melo-
drama (about a wrongdoer punished by other wrongdoers) with all
the explanations removed. It's a very good mysterious tease, and the
dialogue is superb. (Kenneth Tynan said: "Mr. Pinter's ear ranks

with Jenkin's and Van Gogh's among the great ears of history.")
But the play is by no means as genuinely mysterious as *The Caretaker* or as savagely funny as *The Homecoming.*

Yet it's very much better than this film (screenplay by Pinter).
For one thing, the film seems longer, although the original has been
much condensed. (For instance, the theme of Lulu, the seduced girl,
is not completed; she doesn't return after the party, as she does in the
play.) *The Caretaker,* almost the same length originally, is almost un-
cut on film and never comes near sagging.

Some of the difference is, obviously, that *The Caretaker* is better
to begin with. But this new film surely need not have been tedious.
Partly the trouble is Sydney Tafler's performance as Goldberg—
actorish, heavy, and imprecise. (Tafler's gestures are fuzzy and
poorly timed.) Goldberg is a large part, and we don't want constantly
to be conscious of the actor in it. But mostly the trouble is with the
direction—by an American newcomer named William Friedkin.

Friedkin is apparently a scion of Sidney Lumet; he seems to have
studied the latter's film of *View from the Bridge.* Nothing that could
be done simply is left alone. There are lots of shots in mirrors (the
very first shot is in a rear-view car mirror), sudden cuts to high
overhead views, shots through apertures; the camera even leaps forward once to underline a particular word. The screen goes to black-
ness twice (for what are the act breaks of the play), goes into
distorted negative images in a scare scene, goes out of color into black-
and-white in the scenes that take place by flashlight. (As if one
couldn't see colors by flashlight.)

All these lame clevernesses not only distract us from what's going
on; they take time—or at least the distraction makes it seem as if
they're taking time. So the film drags. Friedkin would have done
better to study what Clive Donner did—and did not do—in the
Caretaker film, which is stark, concentrated, and suffocating.

Here's Your Life

(Januray 11, 1969)

THE opening shot (after the titles) is a glimpse—not more than a second—of the hero's face. Then a fast cut to the house in the woods toward which he was turning his head. So small an unconventionality, that opening, yet so radical. Immediately there's a thrill of expectation that this film is in the hands of a director with a vision of his own, and the expectation is quickly confirmed. The film is quiet, visually lovely, done with an ingenuity that enriches its materials instead of distracting us. It's the first feature by a 37-year-old Swede named Jan Troell, a former schoolteacher who has made some shorts for children. Troell directed, photographed, and edited, and he is coauthor of the script (based on a novel by Eyvind Johnson) with the Swedish critic Bengst Forslund.

Here's Your Life it's called, and that's exactly what it's about, in two senses. Here are the formative events of young Olof's life in northern Sweden from ages fourteen to eighteen, during the years of World War I, and here now is life waiting for him at the end of those formative years. The story is quite ordinary: Olof's first jobs, in a lumber camp, in a country movie theater, as projectionist for a traveling movie show, in a railroad yard; his first sex experiences; his first intellectual and Marxist political experiences. But Troell knows that the materials are familiar, knows that his deep engagement with them is apt but insufficient, that he must treat them with art to make them both fresh and endearingly familiar. He succeeds generally well in this, and he does more: he composes a little poem on his country. He is a Swede making a film about Sweden, about the north of Sweden—hard, with men companionable against the cold, with those specially delightful summers that come around in cold regions. It was Lewis Mumford, I think, who said that we can no longer be nationalists but that we must not stop being regionalists. Troell is singing this region through the story of young Olof.

Some outstanding Swedish actors appear in relatively small roles. Per Oscarsson, who was tremendous as the hero of *Hunger,* is a railroad worker. Gunnar Björnstrand, a Bergman stalwart, is the theater owner. Ake Fridell and Ulf Palme and Ulla Sjöblom, all familiar faces if not names, are the touring impresario, a sawyer, a carnival lady of generous virtue. Their parts are all minor, except Miss Sjöblom's, and they are all fine; and the fact that they are *in* those small parts gives the picture an added aura of integrity. Only Eddie Axberg, the Olof, is merely adequate. Troell clearly wanted a credible working-class boy and not a charm bomb, but, with Axberg, credibility is about all he got.

As for what happens to Olof, Troell traces it delicately—with deft, significant ellipses, with humor, and with some lovely photography. Some random samples: a grizzled lumberjack skips along gracefully over logs in a river, and a double image of him appears just behind him, embodying his grace as a separate entity and somehow prefiguring his death. When shy Olof teaches a shy girl how to ride a bike, the sound track repeats the soppy violin-piano music of the movie house—which had been played for a torrid vamp movie. In one unearthly shot, a sled bearing a casket slips along under a sky as completely white as the ground below. And on the old film battlefield of the bed, Troell manages to devise one fresh strategy: the woman and the boy sit on opposite sides dressing in the morning; we see them past the footboard, separated but connected by it.

Not all of the film is successful. There are too many freeze frames —a device that by now has become the June-moon of poetic film making. There is a flashback in color (the rest is black and white) about a minor character's wife, which is overemphasized by the color and is overlong. The character of the carnival woman, though well played, descends from the familiar into the stock. And there seem to be a few wisps and loose ends. (What happened with the railway strike?) The picture has obviously been condensed. The *International Film Guide 1968* says that *Here's Your Life* runs 167 minutes, "the longest Swedish film ever made." The version being shown here runs 110 minutes. Perhaps it was Troell himself who condensed it, and I'm not necessarily regretting it, because a film needs to be a considerable masterwork to justify 167 minutes' running time. But whoever did the condensation left several threads hanging.

Let none of this discourage you from seeing the picture. As is, it's a gentle, humane work by a newcomer of outstanding versatility and gifts. His attitude toward artistic tradition is the one I admire most:

he follows no rule just because it exists, he breaks no rule just because it exists. He comes from somewhere, artistically speaking, but he insists on coming from there, not staying.

Since he finished *Here's Your Life* in 1966, Troell has made a second feature, *Eeny, Meeny, Miny Moe*—with Per Oscarsson as a schoolteacher—which won the Berlin Festival prize in 1968. I hope we can see it soon.

Red Beard

(January 11, 1969)

No picture by Akira Kurosawa can go unnoticed. *Red Beard* was shown at the New York Film Festival in 1965 and, I believe, has since played in the U.S. only at some Japanese-language theaters on the West Coast. Now it is generally released, and there's no mystery about why it was delayed. It runs three hours, and the script is dreadful. Imagine that Frank G. Slaughter had been hired to write a monster installment of a Japanese *Dr. Kildare,* about a nineteenth-century rural hospital run by a gruff old humanitarian doctor nick-named Red Beard whose example alters and inspires a rebellious young intern. Add that Toshiro Mifune plays the older doctor and that Mifune fans demand a scene in which he wins a physical fight against great odds. Here he uses fisticuffs, not swordplay, to disable eight ruffians—or was it eighty?—in a brothel, where he has gone on a house call.

Still I recommend *Red Beard* heartily, though within limits. It may bore the general public, not film specialists. The latter will have the chance to see yet again the mastery of Kurosawa, a man who knows infallibly in every split second of a film (and there are lots and lots of split seconds here) where his camera ought to be looking and how to get it there; how to create an environment through which his narrative runs like a stream through a landscape; how to help his actors—particularly Mifune and Yuzo Kayama as the young doctor—into the very breath of their characters.

The bog is that script. If ever I saw a refutation of the thesis that style is all in art, it's *Red Beard*. Throughout, I was conscious of great talent being lavished on hokum, like Toscanini conducting "The Stars and Stripes Forever" (which at least is short). But, on every score except script selection, makers and students of film can learn wonderfully from it.

Greetings

(*January 18, 1969*)

TWENTY years ago *Greetings* might have been called bright undergraduate humor, which would have been a compliment, but twenty years ago nothing about it could have existed. Not just the topical digs but the style and the very fact that it's a film. I'm glad it does exist because for the most part it's funny.

After approximately 100,000 underground, independent, and student films that aim at satire and miss, it's pure pleasure to come on one that succeeds most of the way and that also justifies its "now" film approach. (Partly by satirizing *that,* too.) *Greetings*—and the title will be recognized by far too many as the first word of the draft-induction notice—is about three young New Yorkers, two of whom come up for induction and all of whom are far out. The story wheels free, winging through such subjects as how to con the draft board, computer dating, porno movies, crackpot solutions to President Kennedy's assassination, and come-as-you-are sex. The big difference between this film and the endless others that flit around such subjects is that Charles Hirsch and Brian De Palma, the scriptwriters, really have wit. They don't assume that they are equipped to write satire simply because they are young, loathe the Vietnam war, like sex, and know some girls who will take off their clothes in front of a camera. More: unlike Robert Downey, who is fairly funny on paper (*Chafed Elbows*), De Palma, who directed, knows something about comic timing—of actors, camera movement, and editing. Considering that he must have worked on a beggarly budget, he gets pretty good results with his color, too.

Greetings is truly and truthfully youthful. *Joanna* is a slick slide over some of the Newest Waves, devoid of personality or credible intent. The script of *Candy* by Buck Henry looks like Henry's attempt to satisfy a producer's idea of mod and yet not lose face with his own swinging friends. *Greetings* is just the work of young people being young, responding acutely but unsententiously to contemporary artistic and social stimuli; not using that response as sole justification, yet insisting on it as the primary reason for the film's being. One of the picture's appeals is the way its air of improvisation conceals its design. Note, for instance, how the scene in the bookshop with the assassination nut leads to the shoplifting girl, which leads to the sequence in which she poses for home movies, which leads to the sequence in which the Vietcong girl poses for the TV camera.

If . . .

(*February 15, 1969*)

LINDSAY Anderson is a sore subject. If there were a way to rate talent in the abstract, he would rank high among the world's directors. He has a fine cinematic intelligence that is fertilized by sensitivity to other arts and by social engagement. But talent is not valuable— to others, at least—outside of works; and Anderson's two feature films to date are serious disappointments. *This Sporting Life* (1963) had some of the best sport sequences and some of the fiercest man-woman quarrels I have seen on film, but it ran out of dramatic energy and ended in debilitated symbolism. His new film, *If* . . ., has passages in it of extraordinary beauty—not only in composition but in concept, editing, and rhythm; yet the dissipation toward the end is even more marked than in his first picture.

The title is from Kipling, I suppose, and the setting is the British public school. The theme is the conflict between the nature of youth and the leather of tradition, with the latter enhanced by the cruelty of older students (called whips) who are given authority and by the beastliness of *all* boys from time to time. Intellectual attitudes toward the public school have changed since the days when Byron grew to love Harrow. (See Orwell's *Such, Such Were the Joys.*) But the very Englishman who recalls his own school experience with some horror rushes to "put down" his son for that same school on the day the son is born. The good schools are still the best manufactories of the right accent; and the right accent is still desirable despite all the prattle about classlessness and all the affections of lower-class or American speech. It is still important to have the right accent to slum *from*.*

The persistence of class devotions among parents and teachers is what underlies Anderson's film. He and his scriptwriter, David Sherwin, begin fascinatingly with a quasi-Brechtian approach: not with the mere use of part titles but with what might be called the view of a compassionate anthropologist. As the boys return to start a new term, Anderson slides a number of "samples" before us, to establish the environment and to place them as personalities, yet not so close as to create a protagonist or organize a story. The purpose is conspectus, and in the first portion of the film, the effect is very exciting as all the cinematic elements are used with great skill by a real mind with a real view.

But doubts begin to assail as quirkiness begins to be evident. First —because it's the first quirk to become apparent—why are there recurrent black-and-white sequences in this well-photographed color picture? No pattern or point is even made of this. Then, why, in a detailedly realistic picture, do grotesqueries of conduct suddenly appear? The padre hides in a huge *drawer* in the headmaster's office in order to hear a boy's apology. The stuffy general keeps orating on Speech Day even though smoke is plainly pouring up from the cellar. What about the loose plot ends? What happened to the two boys who stole a motorcycle? What happened to the whip whose homosexual proclivities were tested by being given a pretty boy as his servant?

More serious than these oddities is the strained imagination that is enforced on the film about halfway through. The first sign is during a tussle between a boy and a café waitress. We suddenly get a glimpse of the pair naked. Whose vision of nakedness is it—his, hers, or the director's? In a later scene the housemaster's plain but curvy wife wanders naked through an empty dormitory. We have seen earlier how, to the virtually isolated boys, she has become a kind of desperate sex image. But where is this naked wandering taking

* English friends advise me that my statements about the "right accent" are out of date. If so, Anderson's implications are anachronistic.

place? Is it real—a demented moment in the sex-starved woman's life? Is it in the imagination of a boy—all the boys? Or is it Anderson's comment? If we knew, the scene in one way or the other would probably be affecting. But no connection with it is possible, and it floats by, its nerve ends disjoined from the rest of the film.

The worst injury, however, comes from Anderson-Sherwin's response to the film's need for dynamic sustenance. About halfway through, I began to wonder what was going to keep this picture going. Up to then it had been varyingly successful "documentation"—of the cruelties of school, the cruelties and beauties of youth. Could it just go on documenting? I wondered. Evidently the director and writer wondered about the same thing, and they inserted plot sequences that wrench the film out of even its previously inconsistent tone, out of its existence. First, there is armed revolt by three boys with blank cartridges; and the film ends in armed revolt with live ammunition. The quad is strewn with corpses. Presumably this is the realization of fantasy, but it fails because it is given no reference of reality or irony. If, for instance, there had been a final shot of the quad *un*littered with corpses, with the rebels meekly marching out of the assembly into the channels of school, it would not have transformed the end into a triumph of the imagination but would at least have located it as bitter dream. Anderson's finish is unsyntactical and ruinous.

All of the cast are competent, with Robert Swann outstanding as a sadistic whip.

On the evidence so far, I would guess that Anderson's best chance to fulfill himself is to find a producer who has keener editorial judgment than his, and to rely on him. What Anderson chooses to do, he usually does extremely well; but his choices in the course of any one film are erratic.

POSTSCRIPT. "We specially saw *Zéro de Conduite* again, before writing started, to give us courage." Thus Anderson in the preface to the published screenplay of *If . . .* by himself and David Sherwin. If only he had taken more than courage, if only he had taken tonal consistency as well. From the first moment of Jean Vigo's film—in the shadowy railroad compartment with steam billowing outside—it is touched with poetic fantasy, so that the final turbulent scene, with the row of dummy figures on the school dais, is the culmination of everything before it. But *If* . . . is also under Brechtian "epic" influence (as Anderson says), and the two elements fight each other.

How I wish he had the editorial judgment I mentioned, or could trust someone who has it, because, in that hypothetical abstract, his talent is large. In fact, for me he is the best directorial talent in British film history.

Stolen Kisses

(February 22, 1969)

I had something of the same experience at François Truffaut's *Stolen Kisses* that I had at Polanski's picture *Rosemary's Baby:* again I realized I had to stop judging the director by his earliest work. Truffaut's *The 400 Blows, Shoot the Piano Player,* and *Jules and Jim* were serious films made in a highly idiosyncratic style. *The Soft Skin* retained that style but showed a curious paralysis of humor, which allowed Truffaut to use trite material quite tritely and to finish with a ridiculous melodramatic ending. *Fahrenheit 451* reminded me in reverse of Disney's hippo trying to be a ballerina: Truffaut was a ballerina trying to be a weight lifter. *The Bride Wore Black* was an arted-up, not-quite-slick thriller. Now with *Stolen Kisses* Truffaut has made a very entertaining romance, full of good feeling but almost devoid of his style, much as if he were stepping deliberately into the general France-for-export business. Polanski began as an investigator of the horror of life and is now a successful maker of horror entertainment. Truffaut's hallmark was lyrical exploration of love and sex, and he has now made a skillful boulevard comedy.

The young hero is discharged from the army, outside Paris, as a psychological misfit. He had been in love, unsuccessfully, and had presumably enlisted because of the girl. Now he pursues her again without success, although she is cordial enough. He has a number of jobs and affairs. At last he wins his girl by quarreling with her and by being busy when she calls. Except for incidental present-day ingredients, the recipe is classic French cuisine. Truffaut skirts dangerously near the stale with such consciously Gallic touches as much of the music and some glory-of-Paris shots (the boy opening his

window on a view of Sacre Coeur); and he even sinks so imaginatively low as to trace a trail of discarded shoes to two lovers in bed.
At no point does the film reach the heights of the best Truffaut
poetry, of which there were traces even in *The Bride Wore Black;*
still there are sequences that would be high points for a lesser man.
The overhead shot when the discharged soldier gets off the bus and
runs across the square. His movement through the empty shop where
he works, following the sound of a song—to discover his boss's
beautiful wife. The youth shaving, chanting into the mirror his own
name and those of his two loves in a kind of manic ritual. The girl's
mysterious pursuer revealed as an admirer, not a private eye. These
scenes are better than mere charm-mongering, which most of the film
does well enough.

The youth is played appealingly by Jean-Pierre Léaud, who was
the boy in *The 400 Blows.* That boy and this youth have the same
name, Antoine Doinel, but it would be pat and mistaken, I think,
to conclude that *Stolen Kisses* is any kind of sequel; the power to give,
or even to withhold, is not what lay ahead of that boy in *The 400
Blows.* Delphine Seyrig, the boss's wife, is exactly the kind of mature
beauty that any youth would dream about—particularly a dream in
which she comes to his bedroom and offers herself to him on condition that they never see each other again. Claude Jade is refreshing
as the girl.

The course of Truffaut's career indicates that he began as an artist
with something to say, said it, and is now left, in his late thirties,
with much executant ability and nothing that he himself really wants
to do; so he will hunt up occasions to use his ability. If this is not all
that his first films promised, he'll probably provide us with lots of
pleasure. The inverse of that sentence is also true: if he provides us
with lots of pleasure, that is not all that his first films promised.

Pierrot le Fou

(February 22, 1969)

JEAN-LUC Godard made *Pierrot le Fou* six features ago (at this writing), far back in 1965. I saw it at the New York Film Festival the

following year and disliked it. It's now released here, and after *La Chinoise* and *Weekend*, I like it even less.

A Parisian TV writer (Jean-Paul Belmondo) is suffocating in his marriage to a rich, pretty, dull Italian girl. (Godard likes to kid foreign accents in French: Jean Seberg in *Breathless*, the Italian girl here.) Belmondo suddenly runs off one night with the baby-sitter, Anna Karina. Complications of murder and torture follow from her previous involvement with gangsters and gunrunners. The pair flee to the Riviera and try to have an idyll; the gangsters follow. It all ends with Belmondo's shooting her, then committing suicide by tying dynamite around his head after painting his face blue. Why waste the resources of color film?

The story would be trite—a mod *Elvira Madigan*—if it asked for any attention as such. It would also be incredible. That this mousy little baby-sitter is also involved with killers and is undisturbed by a corpse in the next room on the night that she and her lover first go to bed—all this would be ludicrous if we were meant to take the narrative seriously. But in a frantic way Godard is deliberately fracturing story logic, using narrative only as a scaffolding for acrobatics, cinematic and metaphysical. The question is whether those acrobatics are consistently amusing and/or enlightening. I think not.

There is the usual Godard barrage of devices, standard even by 1965: verbal-visual puns (VIE in neon turns out to be part of RIVIERA); editing that goes backward, forward, and sideways in time; saturation in film references. Water torture is reprised from *Le Petit Soldat*. There are anticipations: close-ups of comic-strip violence, which prefigure *La Chinoise;* a ghastly auto accident used in *tableau mort,* which prefigures *Weekend*—as does a 360-degree pan when the lovers debark from a small boat. In short, more grist for the movie-buff mills.

For me, the film is a function of three boredoms. (I exclude my own.) The hero is bored by his Parisian life, which precipitates the story. The girl is soon bored by the tranquil island where he takes her, which brings about their deaths. And, principally, Godard is very soon bored. I think that the whole film after they flee the girl's Paris apartment is a series of stratagems to keep Godard himself from falling asleep: improvisations, high-school philosophizing, grotesqueries, and supersanguinary violence. His quick mind seems to have flown ahead to his next film while he is faced with the need to finish this one. Boredom has been a (one may say) vital element in art from Gogol and Musset to Beckett and Ionesco, but in their cases, boredom has been the subject, not the artist's own reaction to the making of his art. The first half of *Weekend,* which is Godard's best work to date, is brilliantly *about* certain boredoms; but in the second half *he* was bored.

His contemporary, Truffaut, seems to be running out of interests and is becoming a body of film-making skills more or less for hire. Godard, a man of larger and more desperate hungers, keeps snatching at themes to nourish his interest. He has gobbled at blood (a midget with scissors in his neck in *Pierrot*), alienation, the Vietnam war, Maoism, fantasy youth revolt, real youth revolt. If anything ever gripped him profoundly, even if only for a couple of months, what a film we might get!

The Prime of Miss Jean Brodie

(March 1, 1969)

No one who admires Muriel Spark's *The Prime of Miss Jean Brodie* can be greatly enthusiastic about Jay Presson Allen's two versions. The best one can do is recognize with a sigh that, as usual, "adaptation" means settling for less. Mrs. Allen's stage version made a wispy attempt to keep the perspectives of the novel; her film version does not. The novel is framed in retrospect. The form is not so obvious as flashback; rather, two planes of time are kept alive—the fate of the women who were once Miss Brodie's pet students and the schooldays that to some extent made that fate. The viewpoint is the author's, the tone is irony, and it is essential to the irony that Miss Brodie never be quite sure which one of her girls betrayed her.

Here the form is all tidied up. Mrs. Allen has taken Mrs. Spark's troublesome time-weavings and straightened them out with carpenter's common sense. Frank Kermode has said:

The suggestion is, in Mrs. Spark's novels, that a genuine relation exists between the forms of fiction and the forms of the world, between the novelist's creation and God's.

Nonsense, replies Mrs. Allen (in effect). A story is a story, and if Mrs. Spark wants to embroider it for book readers, if the form is itself the work's resolution, all right, but she can't be allowed to fool around with film audiences. Fat parts are fat parts, scenes are scenes. The pieces must be assembled, traditionally piled up, and capped with a payoff scene, in which the betrayer tells Miss Brodie of the betrayal and why.

The pity in all this is pertinent, not academic; the form of Mrs. Spark's novel, followed with faithful intention, would have made much better use of the film medium than this rather hack job and would have produced a more subtle and affecting result. This picture has no center; it bears down so heavily on schoolgirlish activities that Mrs. Spark's clean texture becomes rather cute.

Still there is some charm. First is the appeal of novelty. Edinburgh and a girls' school in 1932 are not run-of-the-mill settings; the former is lovely, and the Lallans accents in the latter are winning. Much of Mrs. Spark's diamond dialogue has been kept. Ronald Neame has directed with cozy-matinee competence. And there are two outstanding performances.

Celia Johnson returns to the screen after ten years' absence. In this country she is probably best known for *Brief Encounter* (1946), in which she and Trevor Howard elevated wistful suburban adultery as close as it will ever come to tragedy. I saw her as Masha in a London production of *The Three Sisters* in 1951, and merely to write those words is to feel again her moment of farewell with Vershinin (Ralph Richardson). Now Miss Johnson is back. Wrinkled. When she first appeared in this film, I—and many in the audience—gasped. Her face, like all loved actors' faces, is a calendar. But these calendars tell more than the passing of time; they tell us what time does. Miss Johnson is still beautiful, of course, and now with a beauty free of vanity.

And, of course, she is still a fine actress. Her performance of Miss Mackay, the headmistress, has dignity without pomp and a patient self-reliance even when she seems to be utterly in the wrong. It was a bright stroke of casting to put her in the role. All through the film, sympathy runs for Miss Brodie and against the headmistress. Finally it is clear that, despite her seeming stodginess, Miss Mackay was the sounder of the two. And Miss Johnson's quiet tenacity works retroactively at the end.

As Miss Brodie, Maggie Smith improves through the picture, but she improves as Mrs. Allen's character, not Mrs. Spark's. Miss Smith seems to be playing Jean Brodie's affectations, instead of playing Jean Brodie, a woman with affectations. Mrs. Allen never gives her actress the chance to explore the tension between known and unknown drives that makes the novel's Brodie so terrible; but the film script does provide some "big" scenes that Miss Smith handles with certainty.

Robert Stephens plays the art instructor, Mr. Lloyd—with two arms, instead of the novel's more interesting one-armed war veteran. Stephens is a good actor, but the part has been reduced to a conventional corduroy-jacketed free soul among puritans. The girls in Miss Brodie's set are all nice.

The whole film proves—dare I say this yet again?—that the better a novel is, the less successful an adaptation of it is likely to be. Form and content cannot be easily peeled apart in good works and neatly reassembled. As for the argument that a film made from a novel ought to be judged on its own merits, it applies only when the novel is forgettable; Mrs. Spark's is not. Even the unfortunates who haven't read the book may not be entirely happy with the film; it is unfocused for anyone. The book's admirers will get some reminders of it in somewhat vulgarized form, some glimpses of wonderful Edinburgh, some lightning flashes in Miss Smith's performance, and Celia Johnson.

Bezhin Meadow

(March 1, 1969)

A melancholy half hour comes to us from the Soviet Union—a short made up of still photographs from Serge Eisenstein's lost film *Bezhin Meadow,* accompanied by a voice-over introduction and narration.

By the mid-1930s Eisenstein was in doubtful status with the Soviet regime. Stalin himself had ordered changes in the director's previous film, *The General Line,* saying to Eisenstein, with the wit that endeared him to millions: "Life must prompt you to find the correct end for the film." In 1935 Eisenstein began work on *Bezhin Meadow,* a Turgenev story adapted to present a generational conflict in revolutionary Russia. In the village of the title, a reactionary peasant father finds himself on the opposite side of the struggle from his young son and ends by shooting the boy.

The film was made in the political climate that culminated in the Moscow trials of 1938. When it was finished, revisions were ordered. Eisenstein then worked on a new script with a man whom the sound track of this short film calls "Writer Babel"—the great Isaac Babel, himself soon to disappear in the Stalinist maw. The second version, too, was unsatisfactory to officialdom. The Soviet gobbledygook is thick, but the real reason comes through in Marie Seton's biography of Eisenstein: despite his attempts at strict party obedience, art kept breaking in. In 1937 he published an apologia, *The Mistakes of Bezhin Meadow*. Those of us who sneer have to be very sure that we are heroes.

The narrator of this short film says that the bulk of Eisenstein's picture was destroyed by the German bombing of Mosfilm Studios during the Second World War. But before the war, writing in *Partisan Review* in 1938, Dwight Macdonald cited his sources for saying that the negative had been destroyed, burned for political heresy. The present exhibit seems a small attempt at atonement—and not a moment too soon, to judge by latest reports of changing attitudes towards Stalin's memory.

The short was assembled from surviving scraps by a Soviet director and a film critic. The stills are easily recognizable as the product of Eisenstein's vision and the eye of Edward Tisse, his favorite cameraman. The photographs convey the unique Eisenstein flavor—a distortion of reality that creates higher realism: a combination of masterly screen composition and masterly theatricality. And considering the grim facts that wrap the film, its revolutionary fervor is all the more painful. The very beauties of this salvage job underscore the pathos of what Macdonald has called, with much justice, "the saddest artistic career of our times."

Simon of the Desert

(March 8, 1969)

LUIS Buñuel made *Simon of the Desert* in 1965, a forty-two-minute film about a latter-day religious who emulates Simeon Stylites, the fifth-century saint who spent over thirty years on top of a high pillar, praying and preaching to people below. Buñuel's Simon has spent a number of years on a pillar in the desert and has evidently been a success, in intercessional terms. A rich family is so grateful for his help in getting their prayers answered that they have built a new pillar for him. He is just moving to it at the picture's start. The story chronicles his temptations and fantasies and the efforts that Satan makes, in various guises, to secure his damnation. Satan is at last so impatient with Simon's rectitude that, in the female form which Nick employs throughout, he/she sweeps the saint off to the future in a jet plane and deposits him in a dance-crammed New York discotheque. The ending is tricky and only suggestive, not as completely achieved as the rest of the film; but I can't believe what others have said, that it's an afterthought.

Buñuel was born in Spain (in 1900) and was educated at a Jesuit school. Mother Church had the child and she has made the man, though probably not as she intended. Buñuel, who now lives in Mexico, where *Simon* was produced, has been fighting his birth and breeding all his life, ever since the Eisensteinian bishops in *L'Age d'Or* (1930). But he is certainly not a conventional lapsed Catholic. In the past he has applied the old chestnut to himself—"I am an atheist, thank God"—and he has also affirmed that he belongs to a devoutly Catholic family and that his early education left indelible traces on him. His mockery is that of an intimate, as in the most rigidly religious country I've ever visited, Eire, where the villages are full of anticlericalism. Buñuel, too, spews his sardonic humor on the Church; and also on man; but about God he reserves comment. If a character in Buñuel is introduced as devout, the devotion is usually seen to be true, not hypocritical, even though the person may subsequently change. Nazarin is truly Christian. Viridiana really believes in her vocation at the start of her story. (It's a nice private joke that Silvia Pinal, who was the novice Viridiana, here is Satan in female form.)

Simon is genuine. The tone of the film makes us expect some ultimate exposure of his hypocrisy or failure. No; he is not very bright—he is so keen to bless things that he even blesses a morsel of food he extracts from between his teeth—but he is no faker. The one missing element in *Simon* is also missing from Buñuel's other religious films that I know: the facing of his own contradiction. If he is *not* sure that God does not exist and if he *is* sure that many a priest and parishioner are corrupt, what does that make of God?

Otherwise, this is intelligent fun, tart and compact. Simon, solidly played by Claudio Brook and cleverly written by Buñuel, is amusing in his staunchness. The gratitude of the rich family affects him no more than the ingratitude of the thief whose amputated hands he restores. The malice of a jealous monk doesn't make him feel unworthy any more than the offer of priesthood makes him feel worthy. The film is one of Buñuel's most subtle conceptions, helped greatly because, by its nature, it precludes some of his obsessive, lurid cruelties. No knives this time, no cat pouncing on a mouse, no heads beaten in with rocks; only two items from the Cabinet of Dr. Buñuel: a dwarf—the same one as in *Nazarin*—and the dominant notion of sex as hell.

Buñuel has directed well, calmly getting a sense of movement into a film that is mostly rooted in one spot. There are no flourishes, no arias for director. He plunges right into his story, articulates it cleanly, maintains an engaging but unhurried pace, and quits when he is finished. Gabriel Figueroa, who photographed *Nazarin* and several other Buñuel films, gives *Simon* a smooth gray-and-white tonality that suggests the visual equivalent of Gregorian chant. On the sound track, the various approaches of Satan are accompanied by drum rolls. This reminded me that, at the end of *Nazarin,* when the suffering priest is offered a pineapple by a woman, there are drumbeats on the sound track.

Short and simple *Simon* haunts the memory, possibly because its themes have haunted Buñuel's life and here he has found his most direct way to work with them. We see religion as inescapable complexity for those who once found it lucid. We see prayer, the glory of Christian exercise, made the center of an entire film; and we see prayer as the source both of humble strength and a kind of self-gratification. At the end Simon sits in the discotheque smoking a pipe, physically defeated by Satan but only because Satan has not been able to defeat him spiritually. He is immobilized more effectively than on his pillar but not seduced. Satan has showed him what lies ahead of man, despite all the praying of all the saints, but it doesn't daunt his vocation. I daresay that what Buñuel feels for Simon is compassion; it might even be identification except that Simon has no humor—and except that, like Stephen Dedalus, Buñuel has the "cursed jesuit strain" injected the wrong way.

The Immortal Story

(March 8, 1969)

The Immortal Story is another step in the descent of Orson Welles, a pallid picture in which he also appears but in which he does not appear greatly interested. It runs an hour and was made for television. It shows the marks of TV drama of the fifties, although it was made fairly recently: a predominance of close-ups and two-shots, sofa-vision pacing, an occasional gorgeous prop to give an air of richness but nevertheless a general air of poverty.

Welles adapted the script from an Isak Dinesen story. A hundred years ago, a rich old American merchant is dying in Macao, where he made his fortune. He had heard a story on his way to China fifty years before—about a handsome young sailor picked up and paid five guineas to bed an old man's beautiful bride—and it's now the merchant's geriatric whim to make the story come true. This old man has no bride, but through his clerk, he engages a high-class tart, and then finds a penniless brawny young sailor.

The point is to show the futility of trying to make fiction real, and the story is not a bad Chinese-box device. But Welles's direction misses its presumable aim—Dinesen's timeless, bitter romance—and gives us instead a limp antique. His editing is sometimes bizarre, as in a freak series of cuts when he listens to his clerk read. The only remarkable shot is one that he remembers from his past and keeps repeating here: putting the important person in a scene far in the background and viewing him past a person or object clearly focused in the foreground. The most blatant use of this trick occurs when the clerk leaves the tart in the bedroom and closes the door behind him. She calls; he opens the door again slightly, and she is perfectly framed in the aperture about twenty-five feet away.

The color photography is by Willy Kurant, the only cinematographer other than Raoul Coutard whom Godard has ever used (in *Masculine Feminine*). Here Kurant's work is quite unremarkable; besides, the color registers vary from shot to shot within some sequences. The tart is played by—I might almost say "of course"—Jeanne Moreau. Roger Coggio, an inexplicably noted actor, is the clerk. Some blond young man is the sailor. Welles relaxes through the role of the merchant, wearing his phoniest makeup since *Mr. Arkadin*. He aged more credibly in *Citizen Kane* twenty-eight years ago, but he did a lot of things better then.

I Am Curious (*Yellow*)

(March 15, 1969)

VILGOT Sjöman, the Swedish director of *My Sister, My Love* and *491*, has made a two-part work called *I Am Curious*. Each part is a full-length picture, and instead of calling them Parts One and Two, Sjöman calls the first one *Yellow* and the second *Blue*. These are the colors of the Swedish flag, and the whole work is about Sweden today. Last year when Grove Press tried to import *I Am Curious* (*Yellow*), U.S. Customs seized it as obscene. (In Sweden the film had been passed for showing to anyone fifteen or over.) Grove took legal action against the seizure, and thus—as such actions imply—became a simultaneous plaintiff-defendant: plaintiff because they started the action (the seizure could simply have been accepted), defendant because the government had by implication accused Grove of trying to corrupt American morals.

To skip to the end, the first trial, with a jury, went against Grove. But the U.S. Circuit Court of Appeals reversed this verdict, two to one, and the film is now being shown. The distributors are restricting admission to adults—using their own definition, apparently. The film carries no MPAA rating, so it will automatically be rated X if it plays in any theaters that abide by the MPAA code.

The trouble over *IAC* (*Y*), not specified but understood by all, was not because it shows nudity of both sexes or sexual intercourse —neither of which is novel by now—but because it shows two explicit scenes of oral-genital contact. This *is* new in theatrical films in the U.S. Doubtless Grove had known this and had been prepared for legal action and a test case.

The question of an artist's freedom—a film maker or any other artist—did not arise. It rarely does under law. For all the blather in America and elsewhere about the artist's need for liberty of mind and spirit, the law's concern, insofar as it has any in this area, is to protect the public against the artist. The current legal test for obscenity is threefold. In order to be ruled unobscene, a work must qualify on at least one of these counts: it must be shown not to appeal predominantly to prurient interest; *or* it must be shown not to be without redeeming social value; *or* it must be shown not to offend current community standards. No word about an artist's interest in prurience (which many artists have had) or his disregard of social value (which many have disregarded); certainly no word of the artist's recurrent desire to attack current community standards. Obviously Sjöman's film does not conform to current standards—I mean in films, not in individual behavior. Therefore the case had to be made on one or the other of the first two points.

I was a witness for the plaintiff-defendant. (I'm getting to be a Sjöman court veteran; I was also a witness for *491* in a similar trial.) The gist of my testimony concerned the *wholeness* of the film, an attempt to show that its sexual candor was part of the general candor on all of the many subjects treated. The story is about a young Stockholm girl whose working-class father is a defeated libertarian. (He fought briefly in Spain against Franco.) With some education, she is emerging out of class constriction into an atmosphere of question: questioning everything—social and political and religious acceptances, military tradition, the Bomb, and—inevitably—sexual conventions. Her daring in the last matter is possibly her least adventurous, since her father and his friends are shown as not exactly puritanical.

As a parallel to her curiosity, Sjöman has used an inquisitive method in his directing: moving continually between a documentary style, straight fictional narrative, and even into the making of the fiction. (We often see the crew and the director.) The patent purpose is to show that he is questioning the acceptances of his own world as his central character is questioning hers.

The film seems to me a completely serious work. But that's not much of an esthetic recommendation. To pronounce about a work's seriousness in a review—as against legal testimony—seems almost sophomoric, which may be another comment on differences between art and law. More relevant to criticism, *IAC* (*Y*) is overlong. It could profitably have ended when the girl returns to Stockholm from her country retreat; nothing that happens afterward adds to her experience, in this story. (I hope that the *Blue* sequel opens up some fresh areas.) I'm much less fascinated with the personality of Lena Nyman, the girl, than is Sjöman, and he rests his picture very largely on the fascination of her being, not on her performance. She seems to me porky and stolid and only sulkily interesting from time to time. And Sjöman's directorial methods, which apparently strike him as adventurous, are highly reminiscent of what's been done by other directors in other countries in the last decade. His play with different realities is rather plodding.

What interests me most in this quite honest and quite mediocre picture is its possible effect on concepts of privacy. Putting aside pornography, which is another and complicated subject with its own value system, we might agree that all of human behavior ought ideally to be available to the serious artist. On the other hand, human beings do need areas of privacy in themselves. In order to have a self to communicate, there need to be interior privacies that are *not* communicated, on which the communication is based. From age to age and place to place, those areas of privacy in sexual matters are continually redefined.

But if art is to be deeply affecting, it must speak to the unspoken. Sometimes art accomplishes this by implication, sometimes by explicitness. The new sexual candor in films, of which Sjöman's picture is part, is explicit. It touches privacies. Yet we insist on maintaining some privacies; so it follows that privacy—somewhat different for each of us—must be redefined.

It may be that complete film candor about sexual activity is ahead of us, that intercourse will become as much a material of art in our society as kissing (which is taboo and private in some societies). And, just conceivably, this may be a healthy thing, as the sun temple at

Konarak suggests, because it will mean that privacy would become distinct from prudery: privacy would shift to an area where prudery could not exist.

But I doubt that this will happen in our culture. The mature person who sees explicit scenes like those in *IAC* (*Y*) shoves them back some distance; does not let them affect him erotically; sees them as funny or revelatory of character, not as vicarious emotional involvements. Eroticism is in the suggestive (the best sense of the word), not in the explicit. There is more heat in Bibi Andersson's narrative of an orgy in *Persona,* as she sits in a chair fully clothed, than in all the genitalia of Sjöman's film. The more intrusive a film gets in physicality, the less erotically effective it is likely to be with a mature viewer, who is reluctant to let his most private physical experiences be used as items of reference in a theater.

I'm for Vilgot Sjöman in this matter because, with William Morris, I believe that "no man is good enough to be another man's master" —least of all in morality. But that doesn't mean that all of me is at Sjöman's disposal. So the U.S. Customs can stop worrying about me. If I can't take care of my self, there's nothing they can do to help.

POSTSCRIPT. And then came *I Am Curious* (*Blue*). It explored very little that was new, as against *Yellow.* For much of the time, it simply added different perspectives or information. In plot its chief contribution was to explain how the girl and her lover got the crabs for which they had to be treated in *Yellow.* In technique its chief interest was in how Sjöman had separated out this film from the other, because it's not a sequel, it's a concomitant. It reminded me of something that Akira Kurosawa said when a producer asked him to cut one of his films. "All right," said Kurosawa, "I'll cut it lengthwise." Sjöman seems to have overheard.

Salesman

(April 5, 1969)

WHEN Pontius Pilate asked, "What is truth?" he didn't have to wait for an answer. It's tougher on us. The question persists—we ask it because we breathe—and we're no longer allowed to dismiss it as hopelessly ironic. Answers, attempts at answers, multiply; and nowhere more than in the film.

The use of the camera as purported truth-teller began when film began. From the start, films bifurcated into those propelled by imagination (Méliès) and those bent on showing us the world of fact that we have not really seen (Lumière). The former grew more directly out of previous art traditions, and their modes, at the beginning anyway, were not completely revolutionary. The latter were doing something that had never been done before— still photographs are almost a completely different genus—and nearly from the beginning there has been an evangelical quality to the men who made them. The motions of life around us, they insisted, would reveal truths about that life simply by being captured on film and projected on a screen. The belief has never faltered—one of the most remarkable examples is Dziga Vertov's *The Man with a Movie Camera,* made in the Soviet Union in 1929—and it has surged in the last decade as technical improvements have made filming much more flexible; and also as faith in fictive art has been more severely questioned. Now it's called *cinéma vérité* or direct cinema (Vertov called it Kino-Truth), but it is the same credo that it has always been.

Albert and David Maysles are brothers, film makers devoted to the possible perception of truth through the close observation of fact. In the past twelve years they have spent a lot of time—Albert with shouldered camera, David with portable tape recorder—recording the behavior of various people, as spontaneous as they can get it, then editing the results into film structures. Their best-known works are *Showman,* about the producer Joseph E. Levine, *Marlon Brando,* in which the star turns TV interviewing on its head, and *What's Happening?,* a report on the Beatles' first visit to the U.S. Now they have made their first feature for theaters, *Salesman.* For this they spent six weeks with a team of five door-to-door Bible salesmen, in New England and Florida, and then spent fifteen months editing their footage into a record of several American phenomena, including a dying fall in the career of one of the salesmen.

There are three kinds of episodes in *Salesman:* calls on prospects, scenes in the motels to which the men return at the day's end, and the sales conference (in Chicago). Much of it is fascinating. How could it be otherwise? There isn't a person who passes in the street whose life we wouldn't spy on, at least for a time, if we had the chance. Intrusion into privacy is as human an urge as sex; and it's by no means prurience or itch for scandal that drives us. Somehow some Great Answer may be hidden behind those window shades. If we only *knew* more about others, we could at least be sure that our own insufficiencies aren't unique. A film that allows us to peek is bound to get our attention; and when, like *Salesman,* it also fixes irrefutably some facts about our whole society, it holds that attention longer than it might do otherwise.

Here we get the ugliness of the modern American landscape, what Peter Blake has called "God's Own Junkyard." We enter a number of homes, which seem to have neon-lighted Miracle Miles running right through the living room. (The Maysles brothers have opted to show us only poor homes, although these salesmen call on wealthier prospects, too.) The sales patter beats and repeats against our ears, the phony chumminess with prospects first repels, then numbs with compassion for both seller and sellee. That patter is—surprisingly— Catholic: surprisingly because one expects an American Bible salesman to be Protestant. It's only in the last fifteen years or so that Bibles have been pushed into Catholic homes, so this is the closing of a social time gap.

At night we see the salesmen in their motel rooms, relaxing but not too much. Their straw boss is usually present, and anyway, in front of one another, there are jocular guards to be kept up—no gloating about a good day, no moaning about a poor one. The sales conference in Chicago is, as is the nature of sales conferences, self-satirical.

The picture continues interesting for a good deal of its hour and a half. When it begins to seem repetitious, we forgive it at first because these lives are more incessantly repetitious than most. But this is not life, this is a film; we are not co-workers, we are an audience. Kenneth Burke says: "There is in reality no such general thing as a crescendo." The Maysles brothers are aware of this; so, out of their material, they have quarried the particular story of one of the salesmen, Paul Brennan, and, using the models of fictional narrative, they have tried to give it dramatic structure. But life has not co-operated sufficiently. As drama, the figurative death of this salesman lacks the dimension that it needs to be completely engrossing. There is material missing—of character and conflict and variation—that a good scriptwriter could have supplied; and what we are left with is the consolation that there *was* no scriptwriter, that what we see is spontaneous and unacted.

Almost completely. In a few scenes it seems that voices from other shots—of these men—have been laid on the sound track. And there are indications of the camera's presence in other scenes. For instance, when Brennan comes back after his first bad day, he uses some profanity (the only time in the film). It has an air of bravado, unnatural for him, as if he knew he were being watched and would not be cowed. Some of the other men glance at the camera occasionally. In his car Brennan plays directly to it in the seat next to him. Heisenberg's law has to be trotted out yet again: the fact of observation in itself alters the phenomenon that is observed. The really surprising point is that there are moments—always solitary, always silent—when Brennan seems completely to have forgotten the camera and simply broods. For me, these were the best moments, not only most revealing of him but most supportive of this filming method.

Also—and perhaps this, too, is because the material is given, not made—the film's viewpoint is unclear. Does it mock the commercialization of religion? No. There are some particularly funny lines because of the goods they are selling (Salesman to customer: "Be sure to have the Bible blessed, or you won't get the benefit"), but the patterns would be virtually identical if they were selling encyclopedias or medical reference books. Is the film an indictment of sales as the absolutely central American profession? No, the Maysles brothers have picked scenes that show selling at its grittiest (in poor homes), but the picture's tone is almost as compassionate as mocking. Do they attack selling as corrosive of individuality? No, the picture tries to show a lack of congruence between selling and one man's character. It's a chronicle of this man's failure, not his submersion in sales success. Brennan fails because, quite evidently, he has no histrionism, no con, which three of his fellows have; he has to pump away at it, lamely. Nor does he have the simpleton's sincerity that the fifth man has. *Salesman* is not a criticism of a vocation or the society that produced it. Insofar as it is focused, it's a portrait of a man in the wrong vocation. The others are making out, he is not. And not because, as far as we can see, he is in any way their superior. He is simply not up to this particular mark.

It is this fuzziness of viewpoint and the feeling of plateau that make us feel we are finished with the film before it is over. If "direct cinema" (the Maysles phrase) grabs us immediately with a reality that fiction takes time to manufacture, with the knowledge that we really are there, it has a harder time keeping us there. It lacks the resources that fiction can use to sustain *its* truth: emphasis, distortion, elision, variation, artifice. The most successful direct cinema is, usually, the film about an intrinsically dramatic subject: *Warrendale* (disturbed children), *The Queen* (a transvestite beauty contest), *A Face of War* (Vietnam combat). The daily grind is more difficult.

What is truth? A modern Pilate might say that it's not the monopoly of either fact or fiction film, that life is at least as much of a liar as art, and that if the life is being observed and recorded and rearranged, the line can get fuzzy. Direct cinema is going to play an increasing part in film making because of its ease (not that it's easy to do well but it's easy to *do*) and because it seems like a blow at falseness. But there's one immediate paradox: direct cinema does not cut below facts to truth unless the techniques of fiction are applied. Another paradox: even though "eavesdropping" material is immediately gripping, very soon the content has to feed the basic phenomenon. The snooping into fact, in short, must reveal the content of art. In *Salesman* it does, and in considerable measure, but not enough.

Goodbye, Columbus

(April 12, 1969)

PHILIP Roth's novella has been made into a film that follows the story fairly closely and uses much of the dialogue; yet it ends up as one of those pictures that are superficially faithful and intrinsically false to their original. Producers used to make obvious mincemeat of the novels and plays they filmed; now they are more circumspect. So goes the middle class cultural revolution. Exteriors now have "integrity"; only interiors waver or truckle.

Neil Klugman, lately out of the army, is working in a public library when he meets Brenda Patimkin, a rich Radcliffe girl on her summer vacation. (The locale has been changed from Newark and Jersey suburbs to the Bronx and Westchester.) They have an affair. In the fall he visits her in Boston and finds that she has—probably deliberately—allowed her parents to learn of the affair. Given the conventions of their society, this forces Neil out of Brenda's life. The only alternative would be to force him in—as husband and son-in-law in the business—which he doesn't want. He leaves.

So much thousands of Roth's readers know, and they'll find it all in the film. They'll also find a lot of laughs and a few touching moments. But there are two big flaws. First, the year has been updated from Roth's 1956 to the present. This is ridiculous. The last thirteen years have brought enormous changes in young people, in the quality of their affections and disaffections. Merely to put in some chat about the Pill doesn't update a mid-fifties story that is socially acute in its setting. A small point: Can there still be today a big-university graduate who, like Brenda's brother, collects Kostelanetz and Mantovani records? More important: the incident of the Negro boy who comes into the library to look at Gauguin seems phony today. Not that it couldn't happen; anything *can* happen; but its use here seems coyly humane. (In fact, it's even one of the novella's few weak spots: a slightly strained attempt to give Neil a chance to show compassion, since he doesn't otherwise express much of it.) But the film's anachronism is sealed by Neil's state of mind and the story's climax. The Neil in this 1969 picture is still very much an Eisenhower drifter; nothing that happens in a decade and a half—sit-ins, assassinations, revolts, Vietnam—is expressed or implied in anything he says or does. Yet he is portrayed as barometric to his *Zeitgeist*. Similarly, it's hard to believe that today Brenda would no longer be able to bring Neil home, unaffianced, simply because her parents knew she was sleeping with him. The only way that Roth's story can still hold—as it does hold in print—is as a period piece. A period is a decade nowadays, at the most.

Second, the film is hyperconscious of what Roth takes in his stride: his Jews and their Jewishness. Roth concentrates intensely on what Neil does, sees, hears, and thinks, and, because Neil is Jewish, in a certain time and society, as are most of his friends, a certain American-Jewish society is in the novella. But the film concentrates at least as much on milieu as on character. The producer and director and screenwriter are feeling so courageous at making a noncomplimentary

picture about Jews that they can't restrain their courage. When Neil
goes to the Patimkin house, all the family eat noisily. (Except
Brenda, and why is she an exception?) When the father reaches for
food, the camera is on the table and his hand looms up like a steam
shovel. (Roth tells us that the family heaped and gorged, but Neil's
feeling is not the disgust that this film dinner evokes.) Examples
could be multiplied. No choice in the casting of the peripheral roles,
no reading of line, no framing of an action fails to proclaim that the
Jewish producer and director and screenwriter are pulling no punches.
But Roth didn't punch at all.

The director is Larry Peerce, who made *The Incident,* which unfortunately I didn't see, and *One Potato, Two Potato,* which unfortunately I saw. He is another brightnik who goes to the movies so much
that he hasn't yet had time to become Larry Peerce. His film is satu-
rated with other films. The very first shot—a huge close-up of a
girl's navel (she's in a bikini)—is right out of Godard's *The Married
Woman.* Memories of Truffaut bedeck the falling-in-love scenes.
With Gerald Hirschfeld's unsubtle color camera, Peerce pretties up a
park sequence and supplies the obligatory slow motion when Brenda
runs (almost) naked to the pool. With Ralph Rosenblum's rather
tense assistance, Peerce edits by "linkage" (à la Pudovkin): he cuts
from the steam of an overheated car to the steam of a pot on a stove.
The trouble is that neither the car nor the juxtaposition itself has a
bloody thing to do with the film. He also edits by "collision" (à la
Eisenstein): he cuts from two bodies coming together to a tight
close-up of a red roast of beef on the dinner table. But what for? What
does it do *besides* shock? There is even a reminder of *The Graduate.*
Early in Nichols' film, Anne Bancroft's naked body is reflected briefly
in the glass over her daughter's portrait as she tries to seduce
Benjamin. Here, Peerce's camera follows Brenda from bed to bath-
room until it reaches her father's picture on the bedroom wall, hanging right over a mirror in which Neil can be seen in the bed. Nichols
wanted his touch to be subliminal and was, obviously, willing to let
you miss it rather than pound it. Peerce makes damned sure that you
get it. He shows some ingenuity, but, so far, he's short on taste and
tone.

He does have a feeling for the pace of dialogue. And he handles
the last meeting well, in the Boston hotel bedroom. Here he uses
large profile close-ups of Brenda and Neil for the first time, alternating

on opposite sides of the screen, and the images support the pyramidal feeling of the scene. Then he dollies in slowly on Neil, cutting back to Brenda a couple of times, which heightens the sense of her trying to delay Neil's realization of what she has done.

Richard Benjamin, the Neil, provides more of the performance that I saw in a Broadway comedy called *Star Spangled Girl* (one act) and in a TV comedy series (ten minutes). He has vernacular skill and neat party-patter timing, but he reads all his lines to the same effect: the sensitive youth who uses a defense of sharpness and indifference. The best thing about him is that he creates a certain individuality without coming on strong; the worst thing is that, if you have seen him before, you know he is not playing Neil, he is using the role as a vehicle to deliver more of a situation-comedy formula he has developed.

Ali MacGraw, the Brenda, is a newcomer, a former fashion model. She is pretty and has charm, but I couldn't connect her with her screen family—even in opposition to it. Jack Klugman, the father, could have been good if Peerce were better. Nan Martin, the mother, has the right gold-plated metallic ring.

When the social comment in the film is merely dropped in, it's keen. After a long shot of the flower-stuffed ballroom where Brenda's brother is being married, the camera fixes on the rabbi, with his white-on-white tie, intoning Hebrew. The ancient prayer in a ballroom, along with the flowers and the tie, tell us about one-generation social mobility. And, at this wedding, there is an apparently Gentile photographer wearing a white satin *yarmulke* (skullcap) which enables him to move about during the ceremony. Nothing is made of the matter, so it is effective.

But, for all the good touches, what is missed is what Roth achieved: the telling of an *American* story about American Jews. Essentially, allowing for changes of detail, Neil is a Fitzgerald hero yearning upward toward a golden girl of wealth and "class." *Goodbye, Columbus* even has a good deal of the *Great Gatsby* climate: the swimming, the tennis, the affair that blooms in summer and ends in the melancholy fall. But the film is less like *Gatsby* and more like *Scuba Duba*.

The Red and the White

(April 19, 1969)

OVER the hill a band of cavalrymen gallop to us in slow motion. A dream of war. It's the only slow-motion sequence, this first scene of *The Red and the White,* but it sets the key of the picture: harsh killings, brusque executions, irrational reversals and dominations, all blurring their hard edges into a kind of dream.

Miklós Jancsó is a Hungarian Communist, now forty-eight. All these facts are relevant, together with the fact that he is an exceptionally gifted director. Younger directors of the Eastern European countries are less likely to deal with the establishment of the first Communist state—in this bardic manner, at any rate. And Jancsó has focused on the Hungarians who fought with the Reds against the Whites. His script, which he wrote with two collaborators and filmed in 1967, centers first on a monastery which serves as headquarters in turn for Reds and Whites; then on a White hospital, with some Red patients, which is captured by Reds.

This alternating control—of the monastery, of the hospital, of death —is the theme. Sometimes the power shifts are so fluid that we are momentarily confused about who is who. That, I think, is Jancsó's purpose—moral, not cynical: to show us that, ultimately, it is men who are killing indistinguishable men. There is no preachment (until the very end); the intensity of each officer's belief in what he does, the blind obedience of each soldier, the impersonal juggernaut rolling over men whose own actions helped create the juggernaut that rolls over them—all these make the picture true and numbing.

The sense of history, snatched back and repeated for us, is possible because Jancsó is an epic director, and not in the ad-copy meaning. He disclaims any comparison with Eisenstein, which is wise of him because the connections are clear, although they are filtered through a more modern sensibility. Like Eisenstein, he uses a great many long tracking shots, but he adds the latter-day (Kurosawa) touch of switching a motion to a contrary direction—without cutting—as the interest of other characters takes the scene away. Like Eisenstein, Jancsó relies heavily on the texture of locale: the white odd monastery structures, its galleries and courts; the sedgy bank of the river. (The black-

and-white photography by Tamás Somló is in marvelously composed CinemaScope.) Like Eisenstein, Jancsó brings characters close to our eyes for a few minutes, then lets them slip out of the film: a sparrowlike White general who disposes of prisoners' lives like a fussy old schoolmaster; a Cossack who is prevented from raping a peasant girl, then poses for his execution with affected boredom; a Hungarian Red soldier in his forties who, in a few appearances, conveys patient impotence before he is killed. None of the doomed men, on either side, struggles before he is executed. The effect is not heroism but acceptance of role.

"Come over here" and "Go over there"—these two orders are overused. Prisoners obey these commands, moving to one spot or another, while officers decide what to do with them. The point, I suppose, is to show that commanders must command, even when they are not sure what the command ought to be, but the repetition makes the film drag a bit at times. However, two other sequences that promise to be weary work out quite well. A sex episode seems to start quite arbitrarily, but it is interrupted and cruelly finished in a way that binds it to the film. Then there is a sequence in which White officers make some of the nurses dance for them in the midst of a birch woods. It could have been hellishly symbolic and is not really digested, but at least it is pictorially bearable.

One last demurrer. The final two scenes are so contrary in spirit to the rest of the film that I suspect they were foisted on the picture. (This is a Hungarian-USSR coproduction.) Suddenly the prisoners of history become cinematic heroes, and a small band of Reds march to panoramic death, singing the *Internationale;* soon after, they get a saber salute from a survivor. This ending is not merely disjunctive; the script is supposed to have derived from Babel, and when one thinks of Babel's fate in the Soviet future, this finale is worse than bitter.

But for the most part, *The Red and the White* deals with an ideological struggle in beautiful and humanly contradictory terms. This gives extra dimension to Jancsó's classical, large-scale talent.

Lola Montes

(*May 3, 1969*)

MAX Ophuls' *Lola Montes* was made in 1955, in France and Bavaria, and, except for some festival showings, is now seen here for the first time in unmutilated form. (A butchered, dubbed version was released in 1959.) This is an important event both because of what the film is and is not, and because of what it crystallizes in critical approaches.

Lola Montes was Ophuls' last work; he died in 1957. He was a German Jew, born Max Oppenheimer, who changed his name because his family objected to his becoming an actor. By the time he was twenty-two, in 1924, he had become a theater director and by 1930 is said to have directed almost two hundred productions, including some work at the Burgtheater in Vienna. He began directing films in 1930 and, for obvious reasons, began directing elsewhere in 1932— France, Italy, Holland, Switzerland. By 1941 he was in Hollywood but did not make his first American film until 1947. He did four pictures in the U.S.; probably the best known is *Letter from an Unknown Woman.* He returned to France in 1949 and made four more pictures. Preceding *Lola Montes* were *La Ronde, Le Plaisir,* and *The Earrings of Madame de* . . .

Some critics consider *Lola Montes* to be "the greatest film of all time." To say that I disagree is not merely to quibble with the phrase "all time" as applied to a seventy-five-year-old art; not merely to deplore the facileness with which the accolade of greatness is broadcast in film criticism; it is to differ thoroughly and fundamentally about the means and potentials of film. Some *Lola* lovers concur about some of the flaws I'll describe; but they give different weight to those flaws. That is the heart of the argument.

The film tells the story—a version of it, anyway—of the famous nineteenth-century dancer-courtesan. It begins with the older Lola, playing in a circus in New Orleans. She sits in the center of the ring, as the ringmaster narrates her life, and the bulk of the film is in flashback. We see the end of her affair with Franz Liszt, her (earlier) marriage to her mother's lover, some other embroilments, and her affair with King Ludwig of Bavaria. Throughout, we keep returning to the circus, and it ends there, with people streaming forward to pay a dollar to kiss the hand of Lola, seated in a cage. Thus the structure is cyclical. The cyclical had always appealed to Ophuls: the idea and very title of *La Ronde* (made from Schnitzler's *Reigen*); the reappearance of the earrings in *Madame de . . .*; the recurrence of the lover in *Letter from an Unknown Woman*. In *Lola* the circus ring itself underscores the cyclical motif.

From first moment to last, *Lola Montes* is treasure for the eye, abundant, exciting in its abundance, rich in what Ophuls includes and in the way he handles it. The first things we see are two gorgeous chandeliers descending from a height. (Suggested to Ophuls by the Josefstadt Theater in Vienna?) The chandeliers pass a circus band whose leader is in Uncle Sam costume; and the camera, ever moving, then picks up the ringmaster as he enters. He walks past a multi-leveled swirl of activity to the center of the ring, in front of two parallel lines of girls who proceed to juggle ninepins and to comment in chorus on the tale the ringmaster is telling in flamboyant style. Soon Lola makes her entrance in a gorgeous carriage and is borne to the center. All this to a counterpoint of changing lights and bizarre costumes. (The film is in color and CinemaScope.) The effect—of glittering chaos falling marvelously into order—is precisely the same as in the opera house sequence of *Citizen Kane* and for the same reason, I would guess: both Ophuls and Welles had large theatrical experience. The changes of light within a scene—dimmings, swellings, pinpointings, falls of color—and the knowing use of entrances, these are marks of stage experience.

The most noted hallmark of Ophuls' film style is his moving camera and his cuts from one moving shot to another. Here in the beginning it is used to create a sense of overture, partly by the way the camera grandly ignores the richness of what is happening behind or in passing. The combination of swirl and prodigality promises us largesse: we needn't bother about that dwarf or those splendid horses or that bevy of girls; a great deal is going to spill on the screen.

And it does spill. This is not a matter of purchased Hollywood extravagance. It is Ophuls' gift for selecting the right element of décor, like the low Gothic arch in Liszt's room; for layering every scene with planes of detail ("Details make art," he said) so that the characters are always moving through a world that just happens to tell us something relevant or characteristic about itself at the moment they pass. Examples: the hens roosting all the way up the narrow inn steps; the maimed soldier in Ludwig's castle, past whom the servants have to run when they are on a trifling errand for the king in whose service this man lost his leg; the clown, with whom Lola's doctor waits, who has the voice and demeanor of a prime minister. And, always, these excellent touches are *ignored.*

The visual virtuosity is also in what is done, as well as in the materials included: Lieutenant James chasing Lola crisscross through the descending galleries of the theater; the rope that swings from the stage flies in the foreground as, behind it, Ludwig expresses interest in Lola; the students running toward us down a long ramp to meet Lola's carriage at an angle near the camera; the very last shot, in which the camera pulls back over the hordes advancing on Lola until we are far from her. No fadeout; in the theatrical vein of the film, the curtains close.

All this is superb. There is not a flaw in the *mise en scène,* not a dull frame for the eye. (Well, one reservation: Ophuls either detested or feared CinemaScope and, in some intimate scenes, he puts arbitrary shadows at the edges to narrow the picture.) But after it's all over— *before* then—we are faced with the Chesterton comment. The first time G. K. Chesterton walked down Broadway at night past the flashing electric advertisements, he said, "What a wonderful experience this must be for someone who can't read." In the case of *Lola,* one might add: "Or for those who want to pretend that they can't read."

For the script of *Lola* is just one more teary version of the Prostitute with the Heart of Gold, the whore ennobled by whoring, whom all her friends adore. The matters that made the real Lola an extraordinary woman are omitted completely; we are given only the picture of a woman turned to sexual adventuring by her mother's callousness; who makes her way with her loins; who dramatizes farewells a bit and can develop a little tenderness if the man is a king who gives her a palace; but is only an adventuress, with a touch of Carmen deviltry. To see this Lola as a mythopoeic figure of romance or a figure of the Eternal Feminine, to posit that her story is related to our culture's concepts of romance, is to me a quasi-adolescent insistence on glorifying whores. The difference between, say, Dumas's Marguerite Gauthier and Ophuls' Lola is one between an early attempt to show the particularized humanity of a type and the luxuriant exploitation of the type itself.

The acting of most of the principals is very bad. The late Martine Carol, who is Lola, never could act, and here she doesn't even look pretty. Ophuls spent little time on making her face attractive, even in her younger scenes. Oskar Werner, as her German-student lover, is waxen-faced and cutesy (miscast as a twenty-year-old). Will Quadflieg and Ivan Desny as Liszt and James are sticks. Peter Ustinov, the ringmaster, has merely a fraction of the modulation and shading that he showed in his recent pastry *Hot Millions*. Only Anton Walbrook as Ludwig is substantial.

Some of the *Lola* admirers might agree with all of this; all of them might agree with some of it. Together they reject its relevance. Why? Because they subscribe, with passionate and unquestionable conviction, to a theory of the hierarchy of film values. They believe in selecting and exalting sheerly cinematic values, like the matters I praised earlier, and in subordinating or discounting such matters as those I objected to. To them, this is exultation in the true glory of cinema.

To me, it is a derogation and patronization of cinema. To me, this hierarchy says: "This is what film can do and we mustn't really expect it to do any more, mustn't be disappointed if this is all it does." A chief motive behind the hierarchy is to avoid discussion of the strictured elements forced on film making by the ever-present money men. *Lola* was commissioned as an expensive showcase for Martine Carol. The money men foisted Miss Carol and a cheap novel—by the author of *Caroline Cherie*—on Ophuls, so let's not criticize those elements, let's concentrate on Ophuls' marvelous décor, detail, and camera movement and, by the simple act of appropriate omission, presto, we have a masterpiece.

I disbelieve in this hierarchy. There are money men involved in every art. No one would dream of praising an architect because he designed his interiors well, if he had debased his overall form to please his client's pocketbook. Why a special leniency for film?

Why indeed—in the face of the fact that film has proved it doesn't need it, has achieved *thoroughly* fine work? The worst aspect of this approach is that it crimps the film out of its cultural heritage—the cinematic *and* the literary and theatrical and psychological and social-political—and says to it, "Just go and be cinematic. If anything else is achieved, good. If not, no great matter." It is an esthetic equivalent of the Victorian ethic of "knowing your place."

This concentration on part of a work leads to inflation of the value of that part. Ophuls, who in some ways was masterly, is extolled as a master of romance. To speak only of *Lola,* I see him sheerly as cynic, burdened with this trumpery novel and this mammary star and deciding to give it back to the world in spades. One critic envisions Lola in the circus as a presence "redeeming all men both as a woman and as an artistic creation." *This* woman? *This* artistic creation? The last scene, in which the crowd presses forward to buy kisses of the caged Lola, gave me a vision of Ophuls himself chuckling at the Yahoos who are wonder-struck by this earlier Zsa Zsa Gabor, this "celebrity" in the word's synthetic present-day sense, a crowd scrabbling to pay for a touch of this scandal-sheet goddess. And I also had a concentric vision of Ophuls chuckling at his film audiences, as they press forward to pay for a chance to adulate his caged talent.

Let me give the last word, on this matter of exalting a medium in itself, to the German poet Hans Magnus Enzensberger. Writing about McLuhan in the latest *Partisan Review,* Enzensberger says:

It is all too easy to see why the slogan "The medium is the message" has met with unbounded enthusiasm on the part of the media, since it does away, by a quick fix worthy of a card-sharp, with the question of truth. Whether the message is a lie or not has become irrelevant, since in the light of McLuhanism truth itself resides in the very existence of the medium, no matter what it may convey. . . .

The Loves of Isadora

(May 17, 1969)

KAREL Reisz's film about Isadora Duncan, originally called *Isadora,* opened in Los Angeles in December, to qualify for Academy Award nomination, with a running time of about three hours and with an intermission. After some adverse notice, it was cut to 150 minutes without intermission. It is now released in a 130-minute version with its title amplified to the above. I have seen only this last version. I'm told by a critic who has seen all three versions that the structure has been drastically changed. Whether this was done with Reisz's help, consent, abstention, or disapproval, I cannot say; but I emphasize that my comments pertain only to the third version. Whether those earlier versions were better or worse in total effect, they may have covered some omissions that I note.

The producers of a film usually have control of the film, and often they change it against its director's wishes. When a critic says that "Director Jones has done thus and such," he knows he may be speaking figuratively, that Jones may not be responsible for all defects or all virtues, that the phrase "Director Jones" is a convenience meaning "Jones and those who affected his work." A director protests publicly against alteration at his peril; protest doesn't brighten his chances of future employment.

So here is *The Loves of Isadora,* the work of Karel Reisz *and* the producers Robert and Raymond Hakim *and* some unidentified Universal executives. What hath this conglomerate wrought? On the basis of what is left, the first thing to note is the structural resemblance to *Lola Montes.* This is the story of a celebrated theatrical woman's life with heavy emphasis on her love affairs, told in flashback with frequent returns to the "present," where it ends. Second, the dialogue by Melvyn Bragg, Clive Exton, and Margaret Drabble is of torturous banality. Third, Isadora is largely absent.

No one could gather from this film that Isadora was an important artist ("She was the greatest American gift to the art of dance," said Michel Fokine), a symbol of general cultural forces in explosion and a lasting influence on the dance. The film, as presented, focuses on her men and her egocentricities, with just enough Greek tunics and esthetic asseveration to make her a sort of arty nut, thus justifying to herself and us her bohemian behavior. The result is one more picture about a temperamental star who has lots of lovers—artists and millionaires—and ends up broke. As a dancer, as an artist, this Isadora is about as interesting as Garbo's ballerina in *Grand Hotel.*

What did ancient Greece mean in the life of this Isadora? Well, I think I remember a few fast flashes of some temples in an early montage and a glimpse of the Elgin Marbles in the British Museum. So much for the entire esthetic base of her life. What is shown of her relation to the dance of her time? Nothing. She never sees another dance or meets another dancer. So much—to name just one significant example—for her influence on Fokine. What of her friendship with Duse, her integration with the stirrings of modern consciousness in all the arts? Nothing. Anyone who wants to find out what Isadora was like—both as amorist and artist—would do much better to read the six pages about her in Dos Passos' *The Big Money*.

And anyone who has been trying to cling to some shreds of regard for Reisz had better skip this picture. (I'm speaking only of elements that must be his work.) He made his feature debut in 1960 with *Saturday Night and Sunday Morning,* which was crisp and forceful. Then followed a remake of *Night Must Fall,* so complete a disaster that I couldn't even hold it against him; it seemed more an aberration than a failure. *Morgan!* was just enough of a disaster so that Reisz had to take responsibility for its diffuseness. In what is left of *Isadora,* the scenes as shot—and surely no one else did the shooting—match the banality of the dialogue. All Reisz's concepts are hard-ticket, movie-spectacle clichés. Take, for instance, Isadora's first stage appearance, in a San Francisco saloon: sure enough, we get the stale argument with the reluctant manager and the intercut shots of gaping ruffians with beer mugs. (Those ruffians are possibly the busiest actors in films. Seemingly the same rough faces are overcome by talent and beauty in Arizona and Australia and South Africa. Julie Andrews conquered them last in London's East End in *Star!*)

The clichés thunder on. When the lights fail during a dance recital for soldiers of the new Soviet Union, the goodhearted fellows break into song as Isadora holds a lamp aloft for them. All that differentiates the scene from Jeanette MacDonald solacing the troops is the absence of a soprano obbligato. The *idea* of the scene, from the directorial view, is sheer Romberg. When Isadora dances in Boston, the outrage of the audience is intercut with the fulminations of a street evangelist outside in a style that might have been thought offensively mechanical at RKO in 1935. This sequence is a unit; I doubt that it has been much changed by other hands; and it comes from Reisz, author of the best text on the subject, *The Technique of Film Editing.* There is scarcely a trace anywhere in the film of the intelligence or imagination that might once have been expected of him. Oh, yes, there's a lot of mist in some of the lawn scenes. Possibly it was left over from those *Tom Jones* lawns and just drifted in.

James Fox is brisk and personable as James Fox in a moustache. He's called Gordon Craig, but there's no need to fuss about that. (The script doesn't even get its amorous facts straight; Craig was not Isadora's first lover.) Jason Robards, who is Isadora's millionaire, is miscast yet again in a genteel part and is yet again dull in it. Anywhere but a saloon he is uninteresting. As with Isadora, the emphasis with Sergei Essenin, played by Ivan Tchenko, is on outrageous behavior, with little conviction of the truly great artist at the center, in whom the outrage was peripheral. John Fraser has nice peevish devotion as Isadora's latter-day aide; and there is a funny scene in which Ina De La Haye plays a shocked Russian teacher.

This brings us to Vanessa Redgrave, the Isadora, and the subject of egomania, its curses and blessings. Only an egomaniac could have agreed to show us how Isadora danced—a number of times, too. Miss Redgrave, who cannot really dance at all, boldly attempts to recreate the genius of a great, unique dancer. It's a good deal more hazardous than, say, Richard Burton trying to recreate Edwin Booth's genius in *Prince of Players*.

On the other hand, egomania is probably essential to extraordinary acting talent. If we ignore the contradiction between Miss Redgrave's dancing and what everyone says about it, then what there is of legitimacy in the film comes from her acting. Take such a familiar scene as the one in which they place a newborn baby on the pillow next to the mother. With her freshening imagination, with the uniqueness that she has in *her* art, Miss Redgrave makes the scene new, as if it were the first time we had ever seen it on the screen. When she drives through a tunnel and feels memory and foreboding, the shadow of death really seems to touch her. In her 1927 turban, she moves around the Riviera like an exiled queen with a rag or two left of her court. In these scenes she provides the best job of *voice* aging I have heard since Peggy Ashcroft got old in *Edward, My Son* on Broadway; and, throughout, she has fair success with her American accent.

What we get from her—as she is allowed—is a gawky, beautiful Californian girl, brimming over with *fin de siècle* dedication to Beauty, a term she uses indiscriminately for art and sex. Insofar as this describes Isadora Duncan, which is not very far, Miss Redgrave has created the character.

"She was afraid of nothing; she was a great dancer." Dos Passos' statement is simple, beautiful, complete. There was a film to be made of Isadora's life, with Miss Redgrave: one that did not—could not— show her dancing yet that showed why she was fearless; that showed her as an artistic force; that showed her private life as inevitable for a woman with her hierarchy of values, her extravagances as the frenzy of genius; that showed her as one of the cultural proofs that the twentieth century exists. This long but tiny film—no matter who did what to it—is a mockery of her.

The Round Up

(*May 24, 1969*)

THE success of Miklós Jancsó's *The Red and the White* (1967) has quickly brought us his earlier picture *The Round Up* (1965). What was apparent in the former film is now confirmed: this Hungarian must be ranked among important living directors.

The Round Up deals with the extirpation of revolt. Twenty years after 1848, there are still Hungarian followers of the exiled Kossuth who are making trouble for the Austrian authorities. We see how they are hunted down. The setting—the sole setting—is a stockade in the middle of an immense plain. (Which reminds us that much of Central Europe is flat. Fine battle country.) A number of farmers and sheepherders have been corraled. The authorities know that a rebel leader and his band are among the prisoners but are unable to identify them. The story is a series of tactics to isolate the outlaws. Stool pigeons and murder are involved.

As in his subsequent film, Jancsó has had the help of his cinematographer, Tamás Somló, not merely to put the story on film but to render it in images—to transform the film's concerns into light and shadow and plane and flow of motion. Again the picture is black-and-white, which suits its parabolic intensity, as if color would figuratively have clothed a film that needs figuratively to be naked. Again it is wide-screen, to cope with the immense landscape and the rivulet of history running through it.

History is the key word in considering Jancsó. He has said that his aim in these two films was to clarify the Hungarian past for Hungarians: that he grew up in "the main body of the heroic age of building socialism" (he is forty-eight) and that he shared the "shock when we recognized that, side by side with basically right endeavors, grave mistakes were committed—also by us." Specifically in *The Round*

Up he wanted to smash silly conceptions of outlawry and rebellion, "this phony-romantic conception of the Magyar spirit, this conception which wants to gloss over reality." Yet the picture is not authoritarian; its sympathy is entirely with the rebels. Nor is it cynical and defeatist, although it ends with the authorities' victory. It seems to me realistic: it faces the fact that, given equal determination on both sides, superior force wins, not superior ideals. Gyula Maár writes in the *New Hungarian Quarterly* (Winter 1968):

[*The Round Up*] examines a legend, the romantic legend of the world of outlaws, and breaks it to pieces by showing how it works, the mechanism of the power-enforcement organization and the machinery of oppression.

Jancsó is not trying to disprove the existence of selflessness and heroism. He is trying to show the realistic difference between dying for a belief (which may be a kind of vanity) and helping that belief to prevail, at least for a while.

Cinematically, too, he has applied contemporary perspective to traditional materials. This has been done before, of course: for instance, Mario Monicelli did it with the subject of labor struggles in *The Organizer* (1963). Jancsó works with harsh whites, sudden blacks, lone figures against plaster walls, silhouettes against the stretching immensity of the horizon. The people in these compositions also seem elements in a metaphorical composition—as in Bergman, whom Jancsó claims as one of his masters. This is made even more poignant by the strength of their faces. Even without our contravening Jancsó's purpose, without romanticizing about high cheekbones and fierce noses and fur capes, we can still get from his actors a sense of centripetal ethnic force. His overall structure, too, mitigates against a ballad swell of heroism. Episodes slip into the center of our attention and out again, finished often with a death: and the casual air seems to say that design in life is a function of retrospect.

All through *The Round Up* there are elements that were used later in *The Red and the White:* the white walls, the shock of female nudity in military surroundings (women stripped by captors), prisoners' suicide by leaping from heights, the fiddling of commanders with petty commands. (Also some faces are recognizable from the later picture.) What we are spared here is the hearts-aflame finish of *The Red and the White,* which was a coproduction with the Soviet Union.

This film sticks to its grim guns. But I also disliked the very last shot of *The Round Up*—for quite different reasons. It's a freeze frame: not merely a whacking cliché by now but a stylistic break with the rest of the film, which is in large Eisensteinian curves. This last-second bid for vogue is quite a· different matter from the modernity of concept that has operated throughout.

The Red and the White had its draggy moments, and there are somewhat more in *The Round Up*. Jancsó says he depends very much on his editor, Zoltan Farkas, who is his "first critic." Farkas was more stringent with the later film and, I hope, will grow even more so. It's a matter of trifles all along the way: a second or two here, sometimes a very brief superfluous scene, but these trifles add up distractingly and divert us from the image to the thought of the image makers. Jancsó's work, moment by moment, is more potent than either he or Farkas completely realizes. Less would do more. Less would keep the moving pictures from occasionally lapsing into unmoving picturesqueness.

But, as I noted earlier, Jancsó is one of that now rare breed, the truly epic directors. He is able to depict history and the people snared in it, and without trying to recreate the Battle of Borodino. The two films of his that have been seen here are attempts to accommodate individual psychological verities with basic Marxist views of absolute purpose. Both films have been highly impressive in style and scope. He made a new film in 1968, and he had made four before *The Round Up*. I hope that they are all on their way here.

Midnight Cowboy

(*June 7, 1969*)

JOHN Schlesinger has made a film of James Leo Herlihy's novel (1965), which I reviewed elsewhere. The works are sufficiently alike so that I won't paraphrase, I'll quote some of my review of the book:

Joe Buck is a twenty-five-year-old Southwesterner, an illegitimate child who was brought up by his wayward grandmother. After her death, he makes his way, eventually, to New York in a new cowboy outfit (although he has never been a cowhand), believing that he can earn a fortune studding for rich ladies. But he is not a knowledgeable operator; he is an ignorant, likeable gull. He has never really had a friend and has never really been taught anything. All he confidently knows is the sexual act, and he has believed the myths he has heard about the use he can make of his youthful vigor in the big town. Those myths have filled the vacuum in him of knowledge and ideals.

He arrives in New York to take and, of course, is taken . . . but in the course of all this, he makes the first friend of his life—a runty, crippled Italian-American pickpocket-pimp from the Bronx.

Herlihy's compassionate but relatively unsentimental American-*Candide* style made the book bloom like a flower in the gutter. Schlesinger has understood the book; with intelligence, flourish, and extraordinary skill, he has made an unusually moving film.

His first decision, apparently, was not to tell the story straight. Possibly because he has a visitor's eye, Schlesinger, an Englishman, chose to make a film of present-day American culture with Joe's story as the dominant element in it. The very first sounds we hear are pistol shots and hoofs, and the camera pulls back from a white screen to reveal that it *is* a white screen—in an empty Texas drive-in by day. Then, as a tingling, taking song called "Everybody's Talkin' " comes on strong, we see Joe Buck in his room preening to leave, intercut with the diner and the boss that he is leaving. Soon Joe is on the New York bus. Sharp flashbacks indicate (sketchily) who and why Joe is, and there are also sharp flashes out Joe's window of bare yet garish roadside America. (The Texas of *Hud* comes inescapably to mind and, as if to confirm this, a Paul Newman poster later comes out of Joe's suitcase.) Joe's transistor radio, his electronic rosary, is always in his hands, telling *him* instead of vice versa. As he enters New York, we "see" the faces of women in a street-interview radio show to which he is listening, and soon we are really in the streets. A constant weave of Broadway doorways, weather-and-time signs, neon ads, and thick yellow chunks of taxi blob soon envelops Joe and quickly converts him into some more flotsam in the jetsam.

Schlesinger's sense of pace is so fine that the whirling surface of the film is quite firm; and of course, part of that sense is knowing when to slow down, when to let a scene breathe. In his methods he has at least two hallmarks. One, which I noted in his first feature, *A Kind of*

Loving (1962), is the use of subsidiary action to keep a patch of dialogue from getting static. Example: Joe and his friend, Ratso, are talking in a luncheonette and, at the counter far behind them, a strident woman is telling someone an irrelevant story that we never quite hear but that contributes life to the scene. And Schlesinger likes to begin a sequence with a close-up of some oblique motion that slides us into the center. Example: we follow two drinks on a tray that end up on a sidewalk café table just as Joe and Ratso pass. And the very way that Ratso enters this film is an inspired touch. After the picture is well along, after we have actually forgotten that Dustin Hoffman is one of the stars, the camera slides down a crowded Broadway bar, until it reaches Joe. Then we realize that the person next to Joe is not one more extra, it is Hoffman as Ratso, who strikes up a conversation and launches himself into the story. It's not only a nice twist on Schlesinger's oblique device, it also, figuratively, binds the New York environment closer to Joe.

Schlesinger, an ex-actor, is good with actors. Jon Voight makes his screen debut as Joe, and he is excellent. I've been admiring Voight's theater performances for the past several years. (He was Rodolpho in the off-Broadway production of *View from the Bridge,* for which Hoffman was assistant stage manager and understudy!) Here Voight creates the peculiar innocence of this pubic Parsifal, a man who knows he has set out on an immoral profession but who is completely goodhearted, who reacts directly, even naïvely, to everything. (As compared with Ratso, who "translates" everything coming in his ears and going out his mouth.) There is never a moment's doubt of Joe's reality, principally because of the way Voight uses his eyes: like a child. And he has caught perfectly the Texan speech; listen to the *l*'s when he says "Sally." The film is Joe's drama, and Voight has all the resources to keep him interesting, vulnerable, true.

Except for a short, this is Dustin Hoffman's first screen appearance since *The Graduate.* He has chosen shrewdly, and he plays Ratso shrewdly. There isn't a young character actor around who wouldn't pay for the chance to play the part: a crippled guttersnipe, tricky but winning, who has green teeth and lank dirty hair, and who, to top it all, is dying of consumption. In sum, an actor's dream. Hoffman utilizes all these assets well; he proves again that he is versatile and gifted, no Mike Nichols creation (although Nichols certainly helped him). He has a central vision of Ratso which he has worked out, as he always does, to the last small physical habit. (I saw him again the other night as the crabbed St. Petersburg clerk in the NET production of *Journey of the Fifth Horse* and, besides different makeup and voice, I swear he had different arms.) Here he sometimes uses fairly facile means to get laughs at his commonness, something like a revue-sketch performer playing a New York tough. But the light in his eyes is true weasel light; his discovery of brotherhood is grudgingly real; and the key moments, far beyond the reach of any nightclub mimic, are beautiful. I won't easily forget Hoffman shivering on the cot in their dingy room, saying fearfully to Joe, "Hey, don't get sore . . . but I don't think I can walk any more."

Sylvia Miles gives sexy brass to a call girl who spends a busman's holiday with Joe. Brenda Vaccaro is the career girl who engages him for a night; but instead of acting, she relies on our recognition of hip mannerisms. John McGiver and Barnard Hughes are sound as two unsound older men.

So much of this film is exceptionally good that its uneasy spots are especially troublesome. Ratso's fantasies, visualized for us, mar the viewpoint of the film, which is generally and rightly Joe's; and the fantasies themselves are trite. The Warhol-type party, with Warhol types, looks more like Schlesinger the tourist-shopper than the artist; anyway, can't we have an international statute against trying to depict decadence through wild parties? It never has worked, from *Intolerance* to *La Dolce Vita* to Schlesinger's own *Darling*. Enough, already.

There are touches of overindulgence. A TV remote-control switch lies on the bed where Joe romps with the call girl, and the set switches dizzyingly, with heavy humor. The implied fellatio in a grind-movie balcony doesn't need the sci-fi missile on the screen to make its point. Nor do we need the shots of Joe straining in the sack to please a client. And there is some facile ugliness. Ever since his early documentary *Terminus,* Schlesinger has shown a weakness for the British Free Cinema fallacy: the belief that close-ups of a lot of ugly faces, particularly old ones, prove that (*a*) life is ugly and (*b*) film can tell the Truth about it.

There are a few other flawed moments, like Joe's final brutality with a man whom he robs in order to get the ailing Ratso to Florida. The brutality is in the novel, but Herlihy plays it in the key of Joe's motives, not as a flare of sadism. And the last shot—Joe arriving in Florida on a bus with the dead Ratso next to him, with the Miami skyline superimposed—seems a clever substitute for the good last shot that Schlesinger couldn't think of.

This cleverness, which has obtruded in previous films of his, is still worrisome here; but in *Midnight Cowboy,* his best film to date, there's a great deal besides cleverness, a great deal of good feeling and perception and purposeful dexterity. Films with a homosexual ambience often rely pompously on adduced spirituality: *Boom!* and *Teorema* are two recent miserable examples. But here there is no pomp because there is no argument that love is always present— or always purer—among the wretched and eccentric. This film, which begins with a youth going up to the Big City on a bus and which ends with the changed youth leaving on a bus, simply states that, at any social level, the exchange of trust and devotion is the only sure spiritual oasis; but that, to prejudiced eyes, this seems more incongruous at some levels than at others. By refusing to patronize his characters and by putting them in a realized world, Schlesinger has united us with Joe and Ratso, all of us together hustling amidst the neon but—as we can see—not all of us necessarily lost.

Winning

(June 14, 1969)

THE first shot: in the middle of the vast Panavision-Technicolor screen, a close-up of two flowers, in soft focus. It looks like *Red Desert* revisited. There are distant buzzes on the sound track. The camera moves slowly over a greensward with figures on it, still misty and gentle. Then—*wham!*—we cut to a roadway, the buzzes turn into roars, and cars are whizzing at us. It's a racing picture!

Those opening ten seconds of *Winning* are a sketch of the changes in American culture in the past decade or so. The film proceeds to fill in the sketch, but this opening bit contains the essential ingredients: pop art feeding on high art in order to make the product "smart" for the new pop audience.

I'm not worrying about desecration. Who wants to protect art that can't take care of itself? (Remember those silly protests some years ago against the jazzing up of Bach?) I merely note that a mission has been accomplished: the new, affluent, university-trained middle class has sent its forces into the world of entertainment to wrest pop art from vulgarians and to lacquer it with chic. "We are bright these days," says the new middle, "and those corny vroom-vroom pictures about auto racing will no longer do for people with eight-track stereo and museum memberships. Oh, we still *want* the racing pictures, but please . . . à la mode." The emissaries have done well. They have produced the "adult" Western and the "adult" gangster thriller. *Winning* is an "adult" kids' picture.

It's not the first, but it's a good example—one more fruit of the culture of sententious TV drama and absurdly "sophisticated" photography in *Look* and *Life* and *Esquire* and the Essays in *Time* that soften up important subjects like an Eskimo wife chewing seal-skin for her mate's boots. It's no surprise that James Goldstone, the young director of *Winning,* comes out of television, where he edited and directed. Not only do TV series (when they are made on film) give a man a good technical training, they school him in the essential skill of plucking, from the works of committed men, things that are adaptable and useful and *smart* for show biz. And Howard Rodman, who wrote *Winning,* is another TV alumnus: sharp, agile, frank with all the power of liberated triteness.

Not that *Winning* is a bore. Fast action pictures, if they obey some rough rules of reason, aren't boring. But in its aspirations to be more than an action picture, it is merely modish and intrinsically spurious. Basically it's the same old racetrack story, about the man who loves cars and a girl but who spends so much time with the former that he runs into trouble with the latter. The up-to-date décor includes the fact that he and she sleep together before they are married; that her wavering consists of actually going to bed with another driver; that the hero's mechanic, instead of being an older man like Walter Brennan, is a younger man with—how's this for nitty-gritty?—a hearing aid. And the wife's sixteen-year-old son berates his mother in Hamlet style for her sexual behavior. But under the frank frosting, the old recipe is there. After the hero wins the Indianapolis 500, his mechanic says, "We all made a lot of money yesterday," and he replies, "Jeez, there's got to be more to it than that." True, there's a small twist in that race: the hero doesn't beat his rival (the driver-seducer) in a close finish; the other driver burns out his motor and has to quit. Nevertheless Movieland insists that the seducer go through the Kabuki ritual punishment. After all the fuss is over and he has won the race, the hero socks the bad guy on the jaw.

What keeps the picture from tedium, besides its hard action, is Paul Newman's performance as the racer-hero. Newman simply seems incapable of making a false move or sound. Admittedly, he runs few risks in his roles; unlike the only rival he has in postwar U.S. films, Marlon Brando, Newman rarely hazards much in the roles he chooses. (As does Brando, who so admirably dared to fail in *Reflections in a Golden Eye.*) But whatever Newman does, he consummates. In that ponderous Western *Hombre,* I really believed he was a white man raised by Indians; he *sat* like a stranger. In *Winning* the role is a piece of cake for him, no strain at all. The only moments approaching difficulty are not intense emotional peaks, of which there are few, but the moments of silence, of which there are several. For a small, fine demonstration of imagination quietly at work, look at the scene in which Newman enters the motel room and discovers his wife and friend in bed. He stops, then closes the door behind him, and stands looking at them. Then he turns, opens the door, and leaves. There is a whole spectrum in the pause: of recollection and futility and hurt. Utterly true, utterly free of acting cliché.

Joanne Woodward plays his wife and underplays it with nice timing. Her part is lumpily written. She ascribes her infidelity to a weakness of character that had not been apparent in her. Obviously the plot needed her infidelity and, *ex post facto,* they gave her the quirks.

Goldstone, the director, must be credited with two achievements. He and his editors really provide the sensations of speed, much more than John Frankenheimer did in *Grand Prix,* which I walked out of. I daresay Howard Hawks got real speed into *Red Line 7000* (1965), which I missed; that sort of masculine action is Hawks's forte, as I remember from his previous racing film *The Crowd Roars* (1932). But Goldstone, like Hertz, really put me in the driver's seat.

And out of his big bag of borrowed visual devices, out of Truffaut and Lelouch and others, Goldstone pulls one real accomplishment. Although much of the time these devices are merely impasted modernity, he does use them validly to build Newman's character. The hero needs a certain stillness, a potential for somewhat deeper thought than his fellows, if the film is going to hang together at all. In older days the producers might have given, say, John Garfield a few tinselly poetic utterances to prove that he had a soul. Goldstone uses optical materials—montage and dreamy reminiscence—to suggest thoughtfulness, and it works.

But from the picture's title, with its echo of Henry Green, on through the Antonioni and other derivations and the nervously pared dialogue, *Winning* is a good example of the new hybrid—pop art superficially upgraded. Nine years ago Dwight Macdonald wrote a celebrated essay in which he marked out three cultural areas—High Culture, Mass Culture (*not* folk art), and something he called Midcult. The latter brought High Culture down to bite-size available form. I think the areas are changing; the new culture beavers are bringing Mass Culture up to the middle. As the figurative social masses are being absorbed into the middle class, so Masscult is being absorbed into Midcult, and possibly there will soon be no Masscult left, in Macdonald's terms. I'm of course not talking about such phenomena as the Beatles and post-Beatles rock, which are making their own individual *bona fides* very clear. I mean such traditional Masscult forms as, for instance, the auto racing picture. These are the forms that are being garlanded and "classed up" with the trappings of art. For the best artists, who I daresay are amused by the borrowings, there is presumably only one motto: Onward and Inward.

Last Summer

(July 12, 1969)

Last Summer is about four adolescents at a beach resort. They are almost the only people in the picture, and the performances of the four are the best elements in it. Richard Thomas and Bruce Davison are convincing, particularly the former, but the two girls are extraordinary. Barbara Hershey has something of Jane Fonda about her: the same sexy imperiousness, an attractive yet unfruity voice, quick thrusts of hot true feeling, and sharp timing. Catherine Burns is supposed to be the least attractive of the quartet and, by *Playboy* standards, perhaps she is; but she is lovely in her dignity, in her proud revelations of secret feelings and the childlike humor bursting through her composure. If I were a sultan, I would like to buy Miss Hershey. If I had a daughter, I would like her to be Miss Burns. Frank Perry, who directed, must be credited with good casting and with evoking these performances.

From there on, the compliments get skimpier. The script by the director's wife, Eleanor Perry, adapted from Evan Hunter's novel, is built on a false assumption and on laborious parables. Its false assumption, which it shares with dozens of novels and stories and films, is that, by the very act of choosing adolescence as a subject, it displays special sensitivity, and it invites us to flutter right up there with it onto its assumed poetic plateau. Adolescence has indeed been the subject of some genuinely sensitive films, such as Olmi's *The Sound of Trumpets,* Menzel's *Closely Watched Trains,* Frankenheimer's *All Fall Down* (the family parts), and Troell's *Here's Your Life.* Each of these pictures begins by treating its adolescents as people, as characters, and then achieves sensitivity as an inevitability of the characters' truth. *Last Summer* begins by treating its characters as adolescents and *therefore* sensitive. From the start it has an air of self-conscious idyllic abstraction, which is heightened by the picturesque isolation of sand and sea and sky. All these produced in me an immediate foreboding that the four young people were going to act out a poignant allegory.

They did. The story is built chiefly of three analogous episodes. The first is about a hurt seagull, whose plight unites Miss Hershey and the two boys. The poor bird almost wears a sign around its neck saying "Device." The girl's care for the bird is used to underscore her lonely estrangement from her gallivanting mother and the adult world, and it is not exactly a surprise to us when, after nursing the bird, the girl kills it, shocking and further attracting the two boys.

The second episode is about a Puerto Rican man who comes out from New York because of a prank Miss Hershey plays through a computer dating service. Long before he appears, we know he is going to be sweet and that the youngsters, although they will like him, are going to hurt him, one way or another. He is a figurative gull, as against the literal one.

The third, overarching episode is about Miss Burns. She is the odd girl out. The trio taunt her, befriend her, like her, tease her, and finally injure her grievously, just as with the bird and the Puerto Rican.

Despite this simplistic structure, the moral of the tale is muddied. Does this story tell us what happens to the children of a current generation of sybaritic parents? (Adults are never mentioned except in terms of ridicule or scorn, as mendacious or corrupt.) Or is it about the herd instinct and how the security of group violence overcomes finer instincts? If the former, the case is rigged. If the latter, it's as profound as those Westerns of some years ago in which a stranger rode into a town terrorized by a brute and acted out a little allegory about democracy standing up to fascism.

Apart from Frank Perry's work with the actors, his direction is commonplace at its best, and it's not always at its best. When Miss Hershey and the boys dance, we get a shot of them from the ground up that would have bored Busby Berkeley. When the boys lie on the ground in front of her, their heads fall into a neat pattern. When they smoke pot, we get Indian music on the sound track. When Miss Burns is assaulted by the other three in a woods, we get bird song on the sound track. This irony would be blatant enough, but the bird song is repeated so regularly and mechanically that it sounds as if a phonograph needle had got stuck. The last shot is from a helicopter zooming up and away from one of the boys on the beach. Troell used a vaguely similar shot at the end of *Here's Your Life* because it showed us the countryside out of which the protagonist had come and which he was now leaving; it was a last statement of source. With Perry, it is an empty aeronautical cliché.

The Wild Bunch

(*July 19, 1969*)

THE Western, until quite recently, was especially valuable as a mythic preserve: a form in which Good and Evil could be easily identified and in which Good could triumph. Lately it is becoming an arena for exultation in gore with perhaps a fade-out nod to virtue—a Theater of Cruelty on the cheap. The Italian-made Westerns have done it in an almost childish way: their gore is so patently contrived that the shock is merely annoying, like a boy's thrusting a dead mouse in your face and cackling at your disgust. The well-made American Western is more effective, more lopsidedly truthful.

The Bad Man as protagonist is an old idea in Westerns, but he always used to be Wallace Beery in one form or another, a lovable bandit who reformed or gave his life gladly at the end in expiation. Now we get killers as heroes, and everything we learn about them is intended to make them acceptable *as killers,* not to explain how they went astray and might have been good ranchers or bank tellers if fate had been kinder.

In the hands of Sam Peckinpah, the matter becomes more complex because he is such a gifted director that I don't see how one can avoid using the word "beautiful" about his work. This is not merely a matter of big vistas and stirring gallops and silhouettes against the sky, although Peckinpah understands all about them. It is a matter of the kinetic beauty in the very violence that his film lives and revels in. Peckinpah has been quoted as saying that he included so many horrors in order to make us sick of violence. Well, people will say almost anything in interviews, and it is the tale that must be trusted, not the teller. He is not an oblique puritan, he is a talented maniac who loves his bloody work. And the work is significant.

I missed Peckinpah's *Ride the High Country* (1961), which was reliably praised, and have been trying to catch up with it. I saw *Major Dundee* (1965), about which Peckinpah is said to be unhappy because of interference but which was distinguished by some gritty fighting, some smacking physical detail, and Charlton Heston's most credible performance. *The Wild Bunch,* Peckinpah's latest, runs two hours and forty-five minutes, was written by Peckinpah and Walon Green, and was photographed in Technicolor and Panavision by Lucien Ballard, who was a protégé of von Sternberg and whose distinguished career includes some previous Westerns. This new film is the best Western I can remember since Brando's *One-Eyed Jacks* (1960).

The time is around 1913, the place is both sides of the Texas-Mexico border. The contrast between the U.S. and Mexican cultures is important to Peckinpah (as it was to Brando). It's a handy way to juxtapose Europe and America, to show the pretenses and practices of both, which, under differing delusions and nouns, end up fairly concentrically. The "bunch" of the title is a robber gang led by William Holden; the plot includes two railroad robberies, flight into

Mexico, conflicts with a crooked Mexican general and with Villa's men. From John Ford, Peckinpah has acquired, along with other things, a passion for accurate and revealing Americana, used dramatically. When the behavior gets cinematic, the *look* of the scene is still so genuine that we find patience. The clothes seem to smell of the people who wear them, the tin coffee cups look battered. Peckinpah adds an exotic, somewhat Ophulsian flavor in his décor: the Mexican general has a German military adviser and a bright red automobile. The dialogue is plentifully profane, the frisking with whores is gamy, and both these elements have a retroactive effect: they seem to fill in what has been missing from the laundered Westerns of the past.

Yet there are sentimentalities. When the gang rides out of a Mexican village, which is the home of one of their number, the villagers line the streets singing farewell; and I missed Ward Bond and J. Farrell MacDonald blinking back the tears. Children are used in coarse facsimile of Richard Hughes—as innocent lovers of evil. Sticky symbols occur: a scorpion at the beginning, buzzards at the end. The film is too long; it tries for too much, in incident and theme. (The pronouncements about future Mexican democracy are so blatantly impasted that they don't even taint the film.) Holden is unconvincing as a middle-aged outlaw. The role could have used Henry Fonda or Heston. Holden looks as if he were at a dude ranch, and he can't even walk convincingly. When he and his lieutenant, Ernest Borgnine, stride away from the camera, Borgnine is a man off the saddle, bowlegging along, but Holden is a member of the Screen Actors Guild. Robert Ryan is well cast as an ex-outlaw who leads the bounty hunters who are after the gang, but his usual bile and threat seem a bit thin; he isn't quite brooding enough.

All the rest are good, including Borgnine, Ben Johnson, whom John Ford tried to make a star in *Wagonmaster,* and Edmond O'Brien as a grizzled, chaw-stained old thief. This part would once have been done by Edgar Buchanan, except that here it is written and played as the tail end of a particular life, not as a stock whiskery figure of tarnation fun.

But the faults recede because the violence *is* the film. Those who have complained that there's too much of it might as well complain that there's too much punching in a prizefight: to reduce it would be to make it something else. The violence is in the attitudes as well as the actions. For example, the opening sequence is an ambush by the bounty hunters of the outlaws in the middle of a town, and it doesn't matter a damn to the law or to the outlaws that they are firing through dozens of bystanders. There is a climactic scene in a huge Mexican courtyard which contains so much shooting and stabbing of soldiers, outlaws, girls, and old men—with a wild machine gun clattering away —that it becomes a kind of lethal bacchanalia. It may horrify, but it isn't ludicrous, and this is to Peckinpah's credit.

The details are lurid. When men are shot in the back, blood spurts out—just as it does from disemboweled swordsmen in Kurosawa. A throat is cut before your eyes. And often, when men fall, Peckinpah catches the fall in slow motion. The obvious point of comparison is the end of *Bonnie and Clyde* (there are also other comparisons with that picture), but Penn used slow motion to try to poeticize the deaths of two particular people. With Peckinpah, it is almost always extras who fall in slow motion—depersonalized death. His interest is in the ballet, not the bullet, and this insistent esthetics is perhaps the cruelest of all the film's cruelties. The slow-motion snatches are irritating in two ways: first, because they draw our attention to the film as such; second, because Peckinpah is right—right to remind us that more than one prism of vision is possible at every moment of life and that this prism at this moment magnifies the enjoyment of killing.

The context of the film is the supposed passing of an era and the aging of the outlaws. There are conversations between Holden and Borgnine around the campfire in which they talk about the onset of autumn in their careers, and at the end the theme is well utilized. After a night of carousing with a couple of girls, Holden and Borgnine wake up quietly and quietly go off to pay a debt of honor simply because they're too old to care about saving their skins. It's a well-written and well-directed sequence.

But all that concern with transition and aging is essentially artificial. Peckinpah's real interest is in Western space, men moving in it, men fighting in it. The rest is just the dues he pays, quite skillfully, to form. The overriding effect is of a mania, the eccentric passion of a man who has found a medium that perfectly accommodates his passion. He likes killing, and he does it very well. His art makes it so generic, so tribal, that we can't even go through the usual self-flagellation, American style, about our American violence. That would be too easy, Peckinpah seems to say.

The picture ends in laughter. Ryan's band of bounty hunters are knocked off, so are O'Brien's pals; the two survivors, who were lately enemies, join forces, laughing, using the Mexican peasants' cause as their excuse but knowing it for an excuse. The laughter so closely echoes that of Walter Huston at the end of *The Treasure of the Sierra Madre* that it's almost a quotation, but the ridicule here is not of the vanity of greed but of all absolutes. If the laughter is a bit stagily cynical, that very staginess seems part of the statement.

POSTSCRIPT. Subsequently I saw *Ride the High Country* and was glad of it. The music is studio syrup and the photography by Lucien Ballard is routine, but it is a nice little conventional unconventional Western, and it contains some elements to which Peckinpah has since returned and which he has developed further. There are two leading characters, who are old friends and aging men. The opening sequence is a bit of street Americana into which one of these men rides. There is an attempt—though in strict time, not slow motion—to protract a moment of death. And John Ford's influence is seen not only in the veristic detail but in the use of an eccentric, somewhat overeccentric family of brothers, like the outlaws in *Wagonmaster*. But what is most enjoyable is less any felicity of Peckinpah's direction than the reasonably fresh script by N. B. Stone, Jr.

True Grit

(July 26, 1969)

THE summer rash of big Westerns continues, but at least with some variety. Last week it was *The Wild Bunch,* a powerful picture that harrows the Western form almost to Artaudian depths. This week we are home on the familiar range, yet even here there's an interesting difference.

True Grit is a vehicle for an old star. Readers of the novel may remember it as a book about a girl, but it's a picture about John

Wayne. What's more, although Wayne may make another twenty pictures, this one has an air of valedictory because it marks a change in his persona and relies on his past to certify it, and because it actually ends with a scene of farewell. He plays an aging, paunchy, one-eyed, hard-drinking, profane federal marshal, and inevitably the viewer sees him against the Wayne career of almost forty years in which, give or take a minor characteristic here and there and changes of costume, the star has altered only by very slowly growing older. Audiences have shared that long career with him, and the mutual experience (his and ours) works for him in *True Grit* much like bank interest accrual. Our familiarity with him in one vein is what makes this different vein so effective. The difference here is not in Wayne but in audience acceptances. Nowadays a star is allowed to swear and to get so drunk that he falls off his horse. (Watch closely and you'll see that it's a stunt man who takes the fall.) It's not a performance as was Lee Marvin's drunken old gunny in *Cat Ballou;* Marvin, at the time, had little going for him but his abilities. Wayne has *us* going for him after all this time; and we not only relish his continental body and sprawl as usual, we can relish his age (liberation from the conventions of sex appeal) and his new freedom to break taboos.

Charles Portis' novel was a clever bit of *ersatz* Americana. Although it was short, it was overlong by about a third; still it had a lively sentimental plot: in Arkansas, about 1880, a fourteen-year-old girl hires a federal marshal to track her father's killer into Indian Territory, and she insists on accompanying him. The old-man-and-the-kid bit always works, especially when the old man starts out gruff and the kid wins him over; here the kid is a girl and wins the old man over by physical bravery (his profession), so the project is doubly insured.

The film could hardly have been made with less distinction. This was presumably deliberate. Hal Wallis, the producer, is a man with long experience in turning out moneymakers. He bought a bull's-eye story, got the all-important Wayne, then made sure to engage an old workhorse director who hasn't had a new idea since the beginning of his career—Henry Hathaway. Then they got Lucien Ballard, a gifted cinematographer who can deliver what is wanted. For Peckinpah, in *The Wild Bunch,* he made the very earth look hostile and sardonic. For Hathaway and Wallis, Ballard has produced a series of postcards with unvaried blue mountains on the horizon. Costumes were made not for character but to support audience preconception of character; note the colors in the heroine's clothes. They all aimed at a received fictitious past, which is part of America's truth, just as *The Wild Bunch* was aimed at a veristic past, which is also part of that truth.

But Wallis and Hathaway have bumbled three matters, to keep the film from being as good a homespun contraption as the book. They allowed Elmer Bernstein to pour a thick Straussian gravy over the picture, a score quite foreign to its temper. Second, the chief appeal of Portis' novel was in its prose, supposedly the heroine's own vernacular. The actions themselves have much less flavor than her descriptions of them; and the dialogue, which was amusingly stilted in the girl's narrative, now sounds merely stilted much of the time. Third and biggest bumble, the casting of the heroine, the dullest discovery since the girl in *The Diary of Anne Frank*. Kim Darby's talent need not be discussed; I note only that she doesn't even have the requisite country accent.

A social student can find a few nubs to chew on in this picture. Whenever the heroine has to identify herself or prove her *bona fides,* she does it by stating that her family owns property. Wayne reads us a little lesson in Law and Order; he commands a rat, in mock legalistic terms, to stop eating someone else's food, and when the rat disregards the "writ," Wayne shoots him dead. Only then does Wayne's tolerant cat take any interest in the matter. (The cat, I guess, is supposed to be a liberal.) And when the heroine wants justice, she can't rely on government process; she pays a marshal to help her. All these matters sound quite authentic in context; otherwise they would not be nearly so revealing. They are part of a film made with cinematic conservatism to exalt deeper conservatisms through nostalgia.

Easy Rider

(*August 2, 1969*)

IN 1940 Henry Fonda made *The Grapes of Wrath,* in which, as a dispossessed young man, he travels westward through the Southwest. In 1969 Peter Fonda made *Easy Rider,* in which, as a dispossessed young man, he travels eastward through the Southwest. Fonda *père,* shut out by a greedy and stupid society, finally decides to fight to change that society. Fonda *fils* decides at the beginning, *before* the beginning: he has opted out of society.

Easy Rider is an attempt by some young men at a poetic statement about a world which they feel is inimical to poetry. The declarations of their independence are nomadism, drugs, sex. Fonda and a pal (Dennis Hopper) buy some heroin in Mexico, take it across the border on their motorcycles, sell it in California, then start on their cycles for New Orleans and the Mardi Gras, loaded with money and nothing else. En route they encounter a rancher, a hippie commune, some small-town prejudice against long hair, and an alcoholic thirtyish lawyer who travels part of the way with them. In New Orleans they get two girls in an expensive brothel and spend Mardi Gras with them, including an acid trip in a cemetery. Then the youths meet their fate on the road east to Florida.

All this is very well photographed in color—with a particular eye for the arch of the sky—by Laszlo Kovacs. But the most impressive talent in this film—a potentially important one—is Dennis Hopper. First, he plays Billy to Fonda's Captain America. That's the nickname he uses; Fonda has a flag on the back of his jacket. Clad in old buckskins, Hopper lives in a genial, slightly bewildered "high," present but safe. Second, he is coauthor of the script with Fonda and Terry Southern. Third, he directed. (It's an incidental pleasure of the picture to imagine Hopper switching back and forth between his floating-hippie performance and the rigors of direction.) *Easy Rider* is his first directing job, and it's generally well done, well *seen.* Hopper, who has exhibited paintings and sculpture and photographs, has been busy for some years as a screen actor; for instance, he's the outlaw who gets stabbed by his friend in *True Grit.* To have acted, creditably, a principal role in the first film that he directed more than creditably is an extraordinary achievement. Cheers and hopes for Mr. Hopper.

The best performance—funny, overwhelmingly winning, and quite moving—is by Jack Nicholson as a cynical, hard-drinking young lawyer, drowning in a tiny town, who grabs the back of Fonda's motorcycle going past as if it were a raft that could save his life. There is a crazy sweetness in Nicholson that is pathetic without ever asking for pathos.

Fonda is quiet and tries to work by doing very little, but this kind of stoic film acting—à la Steve McQueen—requires a stronger personality, more menace or protective assurance or sex, than Fonda shows at present. He is presented as a Dharma Bum, a hippie-saint; but the radiations are just not potent enough.

The story is built on some significant assumptions. The first thing we see is the two young men buying and selling heroin, and we are expected to sympathize with them at once. The pot smoking that they do is a matter of course. (The first of the many songs in the score begins "I smoke a lot of grass.") The life style of the two friends, like that of the people in the commune they visit, is taken as understood by us; there are virtually no flashbacks to show us how they got there.

Such analysis as there is in the script gives it its tackiest moments. After the cyclists and the lawyer have fled a small-town luncheonette where rednecks threatened them, the lawyer says, "You know, this used to be a helluva good country. I can't understand what's going on in it." Well, I can't understand *him*. Does he mean that small towns— east, west, north, or south—used to be free of xenophobia and other prejudices? What U.S.A. was that? And when he goes on to explain why money slaves always hate those who are free, the lines take us back to the style of Norman Corwin.

The hippie life is the most credible I have seen in a fiction film. And visually, imaginatively, the picture captures a sense of journey, of the hunger for frontier in American space, the feeling that there ought to be a possible large-spirited life in a large land, that the very ability to *move* through this country is a refutation of the crabbed lives on every side. But, like so many films whose aim is to tell some truth, it gets a bit cross-eyed as it gets closer to its goal. There is some arrant triteness and falseness. The aide-de-camp of the drug buyer, with his dark glasses and death's head cane, is a comic book figure. When the cyclists fix a flat in a rancher's barn, we look past them to the rancher shoeing a horse. (Get it?) The meal with the rancher and his family under the trees is third-rate Steinbeck. The scene where the two youths go swimming with two hippie girls is all cliché; swimming—particularly in the nude—is by now a very weary objective correlative for purity of heart. In the French Quarter brothel there is a glimpse of a whore on crutches, ready for business, that looks like Terry Southern's homage to Max Ophuls. The acid trip in the cemetery is a lot of tedious optical effects without effect. After the pair leave New Orleans, safe with their loot, Hopper gloats that they have "made it," but Fonda shakes his head sadly and says, "We blew it." I don't know what in the world he meant. Why, by any values that they show in the film, had they blown it? It seems a note of pseudo sagacity and rue.

There is some irrelevant chichi in the editing. Toward the end of a sequence, Hopper and his editor, Donn Cambern, give us a couple of splinters of the next sequence before the present one actually ends —brief flashes forward. These flashes are superfluous in narrative terms, and in psychological and esthetic terms they are far from the level of intent for which Resnais used the device in *La Guerre Est Finie*. Especially egregious is a quick flash in the New Orleans section of the fate that lies ahead of the boys; when we see it, it's meaningless, and when we find out what it means, the augury is retrospectively meaningless.

As for that ending, which I had better not reveal, it is a *coup de théâtre* that tries to consummate the sanctification of the two youths, but after the shock is over, it is seen as only a *coup de théâtre*. Which is why I won't describe it. But when a critic can't describe an action for fear of spoiling it for a prospective viewer, that is a pretty fair index of the action's superficiality.

After all these reservations, important as I think them, *Easy Rider* is still an extraordinary first film, particularly from a director who participated in it three ways. It tries to explore its subject, to throw its characters into it and let them take their chances, not merely (in TV style) to substitute new problems for older problems in what is essentially the same script. It tries to create an organism of free flight, and it gets its dynamics from all those interesting assumptions. In cold factual terms, Fonda and Hopper are pretty low types—experienced drug peddlers, criminal vagabonds; but we are given to see two rebels, whose motives are presumably so understandable that they don't even have to be stated, living their own version of two great American romances: the Natural Life (Huck and Tom on motorcycles) and the Frontier. For reasons that have something to do with Jung, a good deal to do with our society's growing unease and at least as much to do with film art, *Easy Rider* often gets what it wants from us.

Putney Swope

(August 30, 1969)

Putney Swope is the most overrated film of the year—and overrated for the most puerile of reasons: it does things that have never been done before in films, anyway in aboveground ones. It's a black comedy (the pun is unavoidable) about blacks taking over a Madison Avenue agency, through a fluke vote something like the election of Corvo's Hadrian VII. Its chief comic method is reversal: blacks turning down white applicants, a wealthy black couple with a lazy blond maid, TV commercials that contain profanity or frank sexual terms, and so on.

Bob Downey, who wrote and directed it, has some wit—as a writer. He showed this in *Chafed Elbows,* which reads funny. What he lacks is film-making ability above low-amateur status, some sense of comic acting and timing, and a ghost of a clue as to any purpose for his wit. *Putney Swope* is just another impudent lampoon of advertising with the switch that it is on film and that the cast has been racially switched. (And it is therefore vaguely patronizing and smug: "Say, here's a new angle on advertising satire. . . .") If I had any faith in Downey's artistic brain, I might say that he was slyly showing us something about the *real* brotherhood of man, but I think that he just had this initial funny idea, that he played along with it until he ran out of gags and gas, then hurriedly cooked up an ending.

Arnold Johnson, who is Swope, plods through with no hint of ability and with a dubbed gravelly voice. There are some beautiful chicks in the picture but not much other talent except a brief bit from Allen Garfield, who was the porno salesman in *Greetings.*

Greetings, which is comparable, was much better than *Putney Swope.* First, it had point and purpose. Second, it had some good performers. Third, although it was pretty untidy, it had more sheer film imagination. Fourth, it was never smug. There is an air throughout Downey's picture of asking for credit simply because it was made.

A further odd point. It's a commonplace by now that the best-produced items on TV are the commercials; more money is spent on them, per minute of running time, than on program material.

(Stanley Kubrick has said that few theatrical films could afford budgets proportionate to what some commercials spend—$100,000 a minute.) Well, in this film lampoon of advertising, by far the best-produced moments are the commercials—those supposedly made by Swope's agency. Besides being the only sections in color, they are the only sections photographed and edited skillfully and performed with some idea of pace. Again I might suspect Downey of making a point of parallelism, if I thought he had sufficient control. But even then it would be quixotic to do an eighty-four-minute film and make seventy minutes of it flabby just for the sake of a caustic parallel.

Staircase

(September 13, 1969)

I had the good luck to see Paul Scofield and Patrick Magee in the London production of *Staircase,* so Charles Dyer's two-character play has for me the aspect of a *pas de deux*—a theatrical excuse to display some fine acting. As a dramatic work, this story of two middle-aged London barbers is a slight exercise in pathos, briskly enough written, which depends for its little life on the fact that the two men are homosexuals. To put it another way, in a play on a less titillating subject, the author's skimpy gifts would be more apparent.

Now Dyer has made a screenplay of *Staircase* and, as screenwriters will, he has "opened it up." The opening only reveals the skimpiness. As a play, it had some prettiness of form: two characters, one place, almost continuous time; the economies were not only Aristotelian, they helped to underscore the isolation of the pair. But now we go upstairs to see Harry's bedridden mum, we go out to parks and graveyards and an old folks' home, a homo pickup is brought in. The shape is smashed. The staircase of the play, under which the two barbers are confined and on which they hear the lecherous feet of their boarder and her lovers, has no meaning. A song called "Staircase," sung by two transvestite performers, is inserted before the credits, to try to give the title meaning. It fails. Worse, the splaying of the script has meant the inclusion of facile ugliness—garbage cans, pee stains

in mum's bed, the gutting of a chicken—all in the cheapest Tony
Richardson tradition.

A good deal of this is presumably the fault of Stanley Donen, the
producer-director. Donen has been an outstanding director of comedies—musicals like *Seven Brides for Seven Brothers* and *Funny Face*
and one of my favorite bits of nonmusical fluff, *The Grass Is Greener*.
In such works he had ease and confident invention. But, venturing into
slice-of-life seriousness, he shows a failure both of invention and of
taste. *Staircase* is a flabby film that, as it goes along, seems to realize
its flabbiness and grabs frantically at bits of sordidness to prove that
it really is serious. No longer a graceful duet, it becomes a waddling
tale, spattered with ugliness, that falls into the biggest sentimental
trap for homosexual material: it pleads for pity.

But it would be a shame if the general failure obscured the one
validity in the film—Richard Burton's performance as the balding
Harry. Burton does not always give good performances, but it is
never for lack of gifts. Sometimes he chooses hopeless material
(*Boom!*), sometimes he is blatantly lazy (*Dr. Faustus, The Taming
of the Shrew*). In *Staircase* he has really concentrated his powers;
Harry has captured and galvanized him, made him provide more
than stunts or surfaces. As the pudgy fearful old poof, Burton could
not rely much on his star persona, and he has not used the vulgar
alternative of being the glamorous Burton at a masquerade. He has
been concerned only with the truth of Harry, the vulnerability and
humility and forlorn gumption, and with the expression of it *as Harry*.
When he tends his disgusting old mother (beautiful Cathleen Nes-
bitt as was), it is Harry's desperate compassion that governs, not
Burton showing us that stars have hearts. When he appears in his
atrocious wig, it is the pitiful man himself, fearful of his pal's comments, not a camp act by the clever Mr. B. There is no crack or
crevice in the entire performance. It's not often that a sex star builds
a role entirely out of what he can perceive and do, rather than out
of what he is or is known as. Harry is an extraordinary work of
character actor's art.

Contrast it with Rex Harrison's performance of the dapper Charlie.
In his own generation (the one before Burton's) Harrison had the
talent and opportunity to become one of its few great actors. If he
lacked the lovely Welsh melancholy that Burton can summon, still
he had a wider range—his own tragic notes, and high comedy that

was superlative. But he lacked one essential. Laurence Olivier has said of himself that he is the greatest actor of his generation, not because he is the most talented—he doesn't think he is—but because he is the bravest. Bravery, of the sort that Olivier means, is precisely what Harrison has not shown in his career. And since *My Fair Lady* —which was mostly a repetition on camera of a previous fine performance—his film acting has shown a continual narrowing, a constriction of spirit, plus an implied reliance on a stature that he hasn't really achieved.

Lately Harrison's acting seems to be floating loose from the centers that once gave it subtle truths. (Remember the exquisite inflections in *Major Barbara* and *The Notorious Gentleman*.) This was bad enough when he was doing things within his usual, though not unvaried, line, like *The Honey Pot*. In the eccentric Charlie, a role that could live only through the sort of creative imagination that Burton shows, Harrison seems quite unable to confront the man with honesty or sympathy. There is no emotional or imaginative connection, there are only surface gestures. This stultifies the moments of anguish (watch Harrison's eyes, which remain dead), but, more shocking, the moments of venom, which should have been easy for him, are mere patter. Most shocking, the speech, intended to be glib for the glib Charlie, is noticeably labored in an actor who used to carve words with swift scalpel strokes.

His performance makes a former admirer very sad. Will he ever recover his powers? Is there time?

Medium Cool

(September 20, 1969)

HASKELL Wexler, one of our best cinematographers (*In the Heat of the Night* and numerous others), has now directed his first feature. He has also written the script and, of course, has done the photography. It's about a TV-news cameraman who comes to understand the radiation effects of the purring bomb in his hands. The title, *Medium Cool*, is evidently a play on McLuhan's phrase "the cool medium." The subject could not be more timely; the visual elements in the film are, naturally, fine; some sequences are exceptionally good; but, ironically, the film itself is trapped in its hero's own pitfall.

The cameraman, played by Robert Forster, works for a station in Chicago, which is apt because the 1968 Democratic Convention can decorate the climax and also because Forster's role is the '68 equivalent of a *Front Page* character—Hildy Johnson with an Arriflex, tough but stoically idealistic. His idealism is in his ruthless professionalism. Then he meets the young widow of a West Virginia miner (killed in Vietnam), who has come to Chicago with her young son. Through his experiences with them, Forster is meant to learn that idealism extends past the ego. And he is also meant to learn that the medium which provides the pleasure in his life is also a medium in which his and other lives are held; and by which, to some degree, they are all fashioned.

The physical feeling of the picture is that it is being pressed close against us. Some of the dialogue has the whip of flexible steel; some of the sequences are small diamonds. There is a scene in which Forster and his girl of the moment, Marianna Hill, whirl around his room naked while she curses him, laughing, a scene that has more sex before they get into bed than most copulation scenes have. There is a minute in the kitchen of that Los Angeles hotel as the cooks work, listening to Bobby Kennedy over a P.A. from the ballroom next door, which ends just as his entourage bursts in through the swinging doors —a touch that suggests early Orson Welles. As Forster walks across the floor of the still-empty Chicago convention hall, they test the recording of "The Star Spangled Banner," rich and thrilling; they stop and start it again—push-button patriotism. Wexler uses the editing device that Richard Lester used so brilliantly in *Petulia*—cutting to parallel yet disparate action or to unexplained flashbacks, eventually weaving the pieces together. Wexler doesn't have Lester's power to convey secret universes that finally meld; still some moments are lovely. And the picture ends with a sardonic reminder—for those who remember it—of the way the Paramount newsreel used to end.

And there are some good performances. Forster, who looks like a young, more fine-grained Jack Palance, has force-in-restraint. Harold Blankenship, who reportedly is an actual Appalachian boy, plays the miner's son with a wealth of proud boyish secrets. Verna Bloom, his mother, has reticent beauty, both in looks and effect.

But, right from the start, *Medium Cool* is infected with the falseness that seems endemic among the new crop of American film truth tellers. The very first sequence: Forster and his soundman shoot a car accident and record the groans of the injured woman on a lonely road. When they get back to their company car, Forster recollects himself and says (the first line of the film): "Better call an ambulance." Phony. It's hard enough to believe that the most hardened newsman would not have called an ambulance before doing his job; but it's impossible to believe that he would have forgotten to call one after he finished—particularly the man this one is later shown to be. The scene is an artificially imposed "meaningful" device, and is followed by others. When the boy, who keeps pigeons on his roof, reads from a book about the fidelity of male and female pigeons, the camera closes heavily on the face of his lonely mother. When Forster and his girl go to a roller derby, the excited crowd yells, "Go! Go! Go!" and the scene dissolves cheaply to Forster and the girl in bed while the sound track keeps yelling "Go!" There is the usual hallmark of the worried realist—facile ugliness: a dwarf attendant at the derby, a badly crippled orderly at a hospital. Wexler wants to include a scene with black militants, so he has Forster invade a black militant meeting, on the trail of an irrelevant human interest story, displaying a professional crassness that would have got him fired or killed long before we meet him. Forster just happens to have been an amateur boxing champion, which helps him to win over his girl's young son. (In *Winning,* Paul Newman had his auto racing to impress Joanne Woodward's son. How does a nonathlete manage in situations like this?) And the last sequence of the picture is a mirror image of the first, so pat and mechanical that it degrades whatever authenticity the picture has been able to establish. Further, that ending is predicted with a bulletin on Forster's car radio, just as the ending of *Easy Rider* was predicted with a film flash. In the latter case, it could possibly be called the character's prescience, but what is Wexler's radio flash? A voice from the spirit world? No, nothing that credible. It is just Wexler refusing to miss a chance for an arty effect.*

Forster's internal drama, which I outlined above, is in fact all in my outline, not in the film. It's assumed that he has gone through

* Others have contended that this incident is Wexler's "homage" to Godard and the ending of *Contempt.* If so, all the worse. Why fracture the credibility of one film to salute another?

what Swedenborg would have called a "vastation," and it's not even assumed by Wexler—we have to assume it *for* him, if the picture is going to hold together at all and make any kind of progress, if the ending is to have even sophomoric bitterness. After Forster is fired by the station (he's told that he broke company rules but the implication is that Big Brother is watching him), all that happens is that he sees a good deal of the widow and the boy, is hired to film the convention, helps the widow to look for her boy, who has disappeared, then runs into the ending. There is some small indication of enlarged humanity in him through his experience with the Appalachians, but it is certainly not crystallized in relation to the theme: the wanton use of the powers of film, especially on television. That theme is simply dropped.

Wexler's inclusion of the convention riots is undigested. What have the riots to do with his theme—*as shown here?* What have they to do with the widow story? The whole episode looks fortuitous, as if the film was being made in Chicago when the riots occurred and as if Wexler just decided to capitalize on them. Capitalization of that sort is a perfectly valid film-making process when the occurrence is really absorbed, but since that doesn't happen here, the riot footage smacks of opportunism. There may well have been a connection between the television age and those youthful protests—as McLuhan has maintained—but it is utterly unestablished here.

At bottom the fault of the picture is that Wexler himself is caught in the same traps as his hero: in the sensual pleasure of shooting film, the ease of creating effects, indulgence in them for their own sake, a kind of reliance on film itself to bail him out of trouble. Of course it's obvious that he's a man of social concerns; the subject demonstrates that. But there were a lot of thirties playwrights who thought that their concern with social truth excused their artistic dishonesty or staleness. Now there is considerable evidence that new American film makers of serious intent and social concern are similarly schizoid: they act as if their intentions and concerns licensed them for the very frauds and clichés and unfulfillments that they would deride in Hollywood hacks.

In Wexler's case it is really painful because he has exceptional talents. He is well worth watching to see whether his future work is free of patent contrivance and glib sensation (in honest causes, of course) or, what is worse, phony candor; whether he will take the trouble to become a thorough artist or will ride along as a flashy, superior Lumet-Frankenheimer clevernik.

Spirits of the Dead

(September 27, 1969)

Go to see *Spirits of the Dead* about an hour after it begins. It's a three-part film—three Poe horror stories made by three different directors. The first two are silly bores, by the justly disregarded Roger Vadim and the greatly overrated Louis Malle. The third is by Federico Fellini. And his horror story is joyous.

Joyous, not because Fellini has no sense of the macabre—after all, his story ends with decapitation—but because he revels in making films and because his darting invention never stops playing around and through the picture, so that even this film of terror plunges us into a sort of Satanic champagne. Fellini's career easily divides into two periods: the first, in which his cinematic mind serves his humanist concerns; the second, in which his humanist concerns are the base for stylist exultation. (*La Dolce Vita* is the transitional film between the two periods.) This short film is very much a matter of execution, not content; although I don't suppose there is a "new" visual concept in it, Fellini's familiar ideas are still exciting.

Toby Dammit, liberally adapted from Poe's "Never Bet the Devil Your Head," is about a sodden English film star (Terence Stamp) as he arrives in Rome to make a Western that will allegorize the myth of Redemption. Stamp, as we can see and the others cannot, is haunted by the devil in the figure of a sly little girl who bounces a white ball. Why Stamp is bedeviled, what the root and resolution are in his little drama—unless Satan is out to keep him from committing sacrilege!—these are never made clear. The film is as hollow as the devil-child's white ball, but how it bounces!

All the latter-day Fellini hallmarks swirl past: the grotesquerie of reality (a street-paving crew who suddenly look like Inquisition torturers) and of exaggeration (faces wearing exaggerated makeup like documents of their past); the flirting fun with the Church (nuns in files and sincerely phony priests); the quick, sharp character touches (as a producer sits down to talk, he runs a comb through his hair); the editing that lets us see the image just long enough for the after-image to touch the next shot; the whirling motion, at the center of which is a smiling mind; the feeling that there is too much, too much, and how wonderful it is to be in the company of a man who has too much.

Sparkling as it is, *Toby Dammit* is not as good as Fellini's previous short film, *The Temptation of Dr. Antonio* in *Boccaccio '70*. Dr. *Antonio* was a simple antimoralistic morality tale, but its very simplicity made the complex style tickle. The new script is a solo for Stamp, with various accompaniments, but it never much engages us as a morality drama and Stamp has little chance to do more than look damned, which he certainly does. But from the first moments of his arrival in Rome (echoing Anita Ekberg's arrival in *La Dolce Vita*) on through a TV interview that made one TV interviewer (myself) wriggle with its truth, to a sort of Italian Oscar-award ceremony, the switch is turned, the motors are humming, Fellini is flying. And if a director is going to concentrate on flash, as he does here, short films are better than long ones, for an obvious reason. I wish the script of *Toby Dammit* were more diabolical, but Fellini's deviltry is almost enough.

The costumes, by Piero Tosi (who did *The Leopard*), are in the tradition of those that Piero Gherardi has done for Fellini, revealing possibilities for contemporary clothes that are quite logical and quite extraordinary. The photography, by Giuseppe Rotunno, is excellent. The editing, by Ruggero Mastroianni (Marcello's brother), is up to the very high level of his past Fellini films. And the score is, as usual, by Nino Rota, whom Fellini calls "a man made of music, an angelic friend."

One point about Fellini's lighting is specially interesting. In his recent films the lighting has been much more theatrical than realistic: low angles, profiles cut out of the dark, the frequent recurrence of silhouettes, and the changes of light during a shot. In *Toby Dammit* an additional theatricality is clear. Often, but especially in the TV interview and in the award ceremony, scenes are lighted like stages and are surrounded by dark, the location in the world is treated like a setting in a theater, and we get the feeling that these lives—by implication, our lives—are being enacted before an unseen audience.

Before whom? Perhaps Fellini has remained more of a Catholic than he likes to admit.

Alice's Restaurant

(September 27, 1969)

ARLO Guthrie's ballad "Alice's Restaurant" is casual, pleasant, funny, and youthfully free, and Arthur Penn's film of the same title tries for the same qualities. Formally, this was a natural move for Penn; in his last film, *Bonnie and Clyde,* he imposed a ballad form on his story, and here he has begun with a ready-made ballad. There is of course a radical difference in the possibilities of this new material, and considering that Penn opted for lyrical and meandering material, which is difficult to sustain in film, he has had some success in recreating its moods.

Guthrie's song, as thousands know, is about his arrest for littering in Stockbridge, Massachusetts, and how his subsequent criminal record turns out to be a boon when he is called up for military induction. Penn, who wrote the script with Venable Herndon, uses this slight narrative as both goal and springboard. We see Guthrie before the incident, traveling the country, spending some time in a college, visiting his dying father (the famous Woody) in a New York hospital, and living with Alice and Ray Brock in Stockbridge. The Brocks, in their thirties, have taken over a deconsecrated church and have made it into a hippie hostel. Alice also runs a restaurant nearby. Arlo's slim story is supplemented by hippie parties, Brock marital troubles, the story of a hippie sculptor on heroin and how he doesn't break the habit, and the closure of the church at the end to form a commune in Vermont.

Penn has ability, and he cannot make a film without giving it some distinctions. There is a lot in this film that is very easy to take. Not much of it will bear close examination, esthetic or otherwise, but a good deal of it—trips (both meanings), comic induction sketches of the Second City or Committee kind, cozy joshing of the police—all these are amiable. Penn is not a director of strong, recognizable character, he is a responder to influences; and he is more interesting here, where he is responding to hippie-ballad influences, than under the influences that produced such grandiose, vacuous responses as *Mickey One* and *The Chase*.

But the fact of his being a visitor on the hippie scene throws light on some things about himself and also about the scene. First, Penn is a great believer in redemption through Beauty. In *Alice* there is a scene of the heroin addict's burial, shot in a snowfall. (The influences are various Japanese directors, together with Fellini groupings.) It's as egregiously Beautiful as anything I've seen since the meeting between the wounded Bonnie and Clyde and the Okies by the stream. Penn evidently thinks that if he just rears back and makes Beauty, it gives a scene validity and conviction. It was not true of the *Bonnie and Clyde* sequence, and the use of Beauty here is even less redemptive because there isn't even a mistaken conviction behind it; it is just aping of the attitudes of the East Village. While the snow falls, Joni Mitchell sings "Songs to Aging Children," as if some special holy innocence were being committed to the earth. But I can't see why this boy's death is any more tragic or martyrish than the drug death of a Puerto Rican slum dweller or a Bowery derelict, to both of whom the hippies might seem stupid freaks. All that we know specifically in this case is that if Alice had slept with this boy more, he might have stayed off drugs. But Penn buys all the hippie tenets of innate superiority and purity without any examination at all.

And carries them further. In *Easy Rider,* made by young men, Peter Fonda is shown as a young sage-martyr. Penn pushes the concept and makes Arlo Guthrie a considerable prig. Every time Arlo gets in trouble with college authorities or police or anyone else because of his outstanding honesty and simplicity, he says something aw-shucks on the sound track about "Seems like I have this habit of gettin' arrested," or some variation of it, as modestly heroic as you please. At one point he works in a nightclub owned by a middle-aged woman. (She knew his daddy in the Movement, for gosh sakes. How old can you *get!*) When she tries to make him, he recoils and prefers to be fired, with a purity that would have done Ruby Keeler proud when Paul Porcasi tried to put his greasy mitts on her back in thirties nightclub flicks. When a mid-teens groupy takes her clothes off to lay him and tells him that she has made it with a number of rock performers, he is nobler than Gary Cooper in refusing to take advantage of her. And since Arlo, played by Arlo, is a pallid personality and is zero as an actor, those scenes are pretty ridiculous.

My own reaction to the hippie views shown in this film is somewhat less adulatory. With some understanding (I hope) of the reasons that hippies exist, I still think they have a considerable tinge of the exploiter about them. One instance: when Alice and Ray renovate their church (no mention, by the way, of where they get the money to free themselves of money society), they and their hip friends line the walls with insulation material. All I could think of was the men working in factories and in sales agencies to make that insulation available to these people who despise factories and salesmanship. Hippies, like all superficial Fourierists, have the effect of mirror-image eighteenth-century aristocrats. Is society constrictive and overorganized? Let 'em eat freedom.

The best performer in the picture is Pat Quinn—new to me—a very sexy young woman, as Alice. The most interesting character is Ray (James Broderick), who is close to forty, apparently, and is hung up on hippies. I wanted to know more about this asynchronous man and how he got that way. It would be glib to suggest an analogy between Ray and Penn himself, but I kept wondering what Penn's feelings were about Ray specifically in this regard. I never found out— nor anything else much about the man.

During the best sequence, the last, I thought of *Walden,* the last chapter. Thoreau tells us that he left the woods for as good reason as he went there. This man who left society because (for one thing) he disliked its routines tells us that he had been in the woods only a week when he noticed that he had already beaten a path, from his door to the pondside. Some such idea may have been in Penn's mind in his conclusion. Alice is in the doorway of the church, which she is soon to leave to found a commune. Loneliness is in the air. As the light changes, the camera moves in slow circles without ever leaving her face: away from her and back to her in slow circles. It is the moment in the film that seems to me most authentically Penn's; much of the rest is just more or less "with-it."

A Recent Change

(*October 4, 1969*)

A new book about films of the forties, Barbara Deming's *Running Away from Myself,* underscores by implication a significant change in films of the late sixties. Miss Deming reminds us that American films in the past usually reflected the truth of American lives without intending to, that films manufactured as commercial entertainments were inescapably the products of contemporary psyches. Within the last two years, there has been a considerable shift: many American films have quite consciously come to grips with social phenomena and with psychical states. (I'm speaking of a change in well-budgeted theatrical films. Underground films have always tried to treat those matters; one of their reasons for existence was to compensate for the lack of such honest encounter in aboveground films.) This is not to say that psychical substructure has disappeared from these new films, but much that was once implied, or that seeped into films only because it couldn't be kept out, is now there by explicit design.

The press has been filled lately with accounts of the arrival of the New American Film. Several writers now give the cause of the change as the appearance of *The Graduate.* When I said in late 1967 and again in early 1968 that *The Graduate* was "a milestone in American film history," the remark was greeted with some hoots. I raise this matter now not (or not only!) to show my acuteness but to emphasize how quickly the flood has burst, as if it had been straining at a dam that suddenly gave way. Obviously some of these films were in the works before *The Graduate* appeared and obviously some "personal" films were made in this country before then, but there is now a strong new direction of which the Mike Nichols picture was the first visible marker, which is what "milestone" means. Let's define a "personal" film (yet again) as one made primarily because the maker wants to make it, not as a contract job: analogous—as far as the conditions of the medium permit—with a poet's writing a poem or a sculptor's making a sculpture. In most of these new films the subject is some aspect of American society or some experience of the film maker's that he wants to investigate and correlate with the world. In more of these films than is usual, the director wrote or collaborated on the script.

Here are some of those films, with brief comments on the ones I have not previously reviewed: *Greetings; Goodbye, Columbus; Last Summer; Easy Rider; The Wild Bunch; Putney Swope;* and *Medium Cool. The Learning Tree,* written and directed and scored by Gordon Parks, is banal and sentimental, with glucose music, and is photographed through assorted fruit juices, still it's an important event in our cultural history. Autobiographical films, such as this, are rare enough, and this one is by a black man about his boyhood. Ten years ago it would have been inconceivable that such a film would be financed by a major producer; five years ago it would have been highly improbable. And (as I have seen) its effect on a black audience, after decades of peripheral black characters in films, is one of powerful affirmation. *The Rain People* is a tedious and affected piece about a wandering young wife who "finds" herself through her experiences with a simple brute, the most painful bit of preciosity since Jack Garfein's *Something Wild* (1961), which it resembles somewhat; but it was written and directed by Francis Ford Coppola, a relatively recent film-school graduate, and whatever it is, is his own work. *Who's That Knocking at My Door?* is by a more recent film-school graduate, Martin Scorsese, and, on a more limited budget, again tries to substitute camera shenanigans for insight and content. The script, about Italian-American puritanism, is like that familiar first novel or first play where the author thinks that, because he's writing of what he knows, we will automatically want to know it, too; but distinctly it is a "personal" film.

To this list could be added *Midnight Cowboy,* which was directed by an Englishman but is American in backing, source, script, and performance, and *Alice's Restaurant.* Arthur Penn is so well established that he can do virtually anything he chooses; his choice of *Alice* reflects the temper of the past two years.

All these films are of widely varying quality and prove yet again that to make a film "personally" is no guarantee of artistic success. A straight commercial film like *Hot Millions* or *Funny Girl* is a lot more rewarding than *The Rain People.* And as I've noted in my reviews of *Midnight Cowboy* and *Easy Rider* and *Medium Cool,* the free souls have their own falsities and self-indulgences to beware of. Still the promise in this new situation cannot be denied.

The reason for this change in the theatrical mainstream is, I believe, the presence in the audience of the millions who have been pouring out of colleges since World War II and the millions who are now in colleges: their growing interest in the film, their *reliance* on it as they rely on no other art; their rejection of the ludicrousness of the commercial formulas or at least their refusal to accept them as the totality of film; their concern with the society in which they live and themselves in that society; their shame at the difference between what has been happening in the best postwar European film and what had been happening here. The fulcrum on which the change turned, the essential component, is of course the mind of the financier. He has seen where the money is; at least he has seen that the money is not unfailingly where it used to be; and in his bewildered thrashings about, he has sometimes thrashed toward the personal film. Of course if personal films don't make money, their future is going to be in difficulty, just as the French New Wave ebbed quickly when so many of those films lost money; so I hope that, in his floundering, the financier doesn't back too many fuddlers or fashionmongers or arty fakers, or else he will soon do his floundering in some other direction. But in the meanwhile a needed avenue has been opened.

(October 18, 1969)

Two weeks ago I commented on the freshet of "personal" American films and said that I thought the chief reason for it was the changed attitude toward film of the millions who have come out of college since the Second War, along with the millions who are now in college. Some friends, and some others, have pointed out that this influence is far from discriminating, that young people tend to like films about young people so long as their lives and language and viewpoints are credibly portrayed. This is true enough. In speaking to young audiences, I often run into difficulties in criticizing one of "their" pictures. If it tells the "truth" about their lives, they don't care much about other factors, just as many blacks don't care about the faults in *The Learning Tree* because at last there is a full-dress American picture about blacks.

Well, I do care about those other factors—very much; and I think they are important to the future of all films, including those about young people and about blacks. But need a blessing be unmixed in order to be a blessing? In the world at large, has one endorsed the mindlessness of much youthful protest just because one is glad of the true moral passion in some of it? The fact is that if this new generation (and it runs up to forty) did not exist, evasive and shortsighted as it often is, this new movement in American films would not exist. They have at least budged the American film industry off its upholstered seat. What needs to happen now is for young people—white and black—to get over the first rush of gratitude, to get bored with films that *merely* tell some truth about them. By their own discrimination, they can help prevent the industry from slickly pandering to them; and thus keep alive this new, but vulnerable, chance for growth.

The Bed Sitting Room

(October 18, 1969)

PLENTY of voices have been eager to tell us that the trouble with Richard Lester's new film is that it is like *How I Won the War*. I disagree. The trouble with *The Bed Sitting Room* is that it is different.

The new film was made from a play by Spike Milligan and John Antrobus that had a considerable London run. Milligan has a big British reputation, but on the basis of this work and of his performance (he is in the film), the reason eludes me. Lester knows him from early television days and evidently values him. Some auld acquaintance should be forgot.

This is an absurdist farce that takes place three years after the next war, which itself lasted two minutes and twenty-eight seconds. London, where it all happens, is now one vast cindery waste, with a few scraps of buildings, the top of St. Paul's dome, a park of broken dishes, and a mountain of shoes. Underneath runs the underground, pointlessly, powered by a bicycle-driven generator and inhabited by a few of the twenty people left. Overhead floats an orange balloon containing two detectives, ordering everyone (such as that is) to keep moving, so as not to present a good target to a potential enemy.

In this nowhere that is left, while we hear a series of old-timey foxtrots and waltzes on the sound track, we get a series of farce and slapstick situations based on fantasy extensions of atomic mutation. A lord (Ralph Richardson) feels he is changing, and does change, into a bed-sitting room. Rita Tushingham, impregnated in the underground, gives birth after eighteen months to a monster, which is followed soon by a normal baby. Her father changes into a parrot, her mother into a wardrobe. It all ends with qualified hope—or perhaps it's mitigated despair.

At least two elements in this film are superb. David Watkin's color photography insists on every delicate shade of the waste on which we look, as if to satirize the idea of beauty here by rendering desolation with artistry. And most of the acting is excellent. Michael Hordern provides more of his fine flustering as a doctor; Arthur Lowe and Mona Washbourne, the girl's parents, give several textures to everything they do; and the joy above all is Ralph Richardson. He passed through a long period in his career when he indulged in the most eccentric line readings since the ripest Orson Welles. He was never dull, but he was often mannered and self-serving. Then, with his Peter Teazle in *The School for Scandal* on Broadway a few years ago, a more mature Richardson appeared, interested only in the work in hand. His crazy lord here is a gem of caricature. You will go far to hear the English language inflected with more comic relish and subtlety.

Lester's direction suffers from the constrictions of a theater piece, as it also did in *A Funny Thing Happened on the Way to the Forum* and as it did not in *The Knack*. In the latter he and his adapter blew the play apart and mixed the bits with other bits of their own invention to make himself at home cinematically. *The Bed Sitting Room* has doubtless gone through changes from the original, but it remains a linear work which Lester follows in two-dimensional fashion. If the incidents had some interest or humor, which they generally do not, that would help. If the dialogue were funny, instead of being laden with weak puns and fizzled firecrackers, that would help, too. Even in *Forum* Lester made some attempt to break loose and release his proper gifts. Here he just plods along obediently after a gag writer's script.

Thematically, the film is weak. The idea of showing nuclear horror obliquely, through farce, is more stimulating than the earnest confrontation with holocaust that some have tried; at least it involves an attitude, a transformation, that makes it more real than piled-on screams and flames. But the very first view of Hordern in his nonexistent surgery or of the family living in the subway car tells us absolutely everything that can be done in this vein; the rest is protraction, not development. And what if it were developed—what would the point be? A warning? To us? The people who elected Richard Nixon—or possibly would have preferred Hubert Humphrey? Basically what puts me off this film is the implication that it contains a learnable lesson—a lesson that we could all easily master, if only we would, like quitting cigarettes.

Not *How I Won the War*. The really extraordinary qualities of that film are that it keeps its ridiculousness and its killing all in the same unfunny comic vein; and that Lester's pyrotechnics, with this base, cut through the pietism in all of us to some truth: namely, we doom ourselves and *we like it that way*. If Freud's thesis of the death instinct is now questioned, still it is obviously and demonstrably true that there are things we are more afraid of than death, such as losing the perquisites of ego.

Last winter I showed *How I Won the War* at a Midwest university and spoke about it. Afterward a young man told me that the picture had really frightened him because he recognized, in the militaristic hullabaloo, the very feelings that had filled him when he marched in an anti-Vietnam protest a few days before. That reaction helps to clarify why *The Bed Sitting Room* is a weak humanist gesture and the earlier film is an important work. *How I Won the War* tells us (with breathtaking art) not that we are victims of bureaucracies or that enlisted men are angels slaughtered by devilish officers but that we are *all* finished—now that we have the power to finish ourselves —unless we go deeper than deploring war and nuclear bombs; unless we see that even the people who say that war is horrible—and say it quite sincerely—are part of the horror, too.

As for Lester and his professional future, if he had a debt to Spike Milligan, he has now paid it; and can move on.

Adalen 31

(October 25, 1969)

Bo Widerberg, the Swedish director, has now restated his idea of tragedy: make everything very pretty and then have something nasty happen. In *Elvira Madigan* honey flowed all over the screen until the two lovers met their suicide pact. Calendar pictures of perfect idylls fluttered past until harsh facts—perfectly predictable except to that dumb pair—ripped the pictures across. Now we get *Adalen 31*, which takes place in the Swedish town of that name in 1931 and is about an actual labor struggle in which five people were killed. (The event led to the victory of the Social Democrats, who have dominated Swedish politics since then.) Before the fight we see the family of a striking worker, all of them perfectly happy; and we see the boss's family, every one of whom is a perfect doll. The worker's oldest son meets the boss's beautiful daughter, and they make beautiful first love. (Including a pregnancy, which the girl's mother handles with beautiful aplomb.) Then, into all this beautifulness, the harsh confrontation irrupts. But even that trouble—like the *Madigan* suicides —is beautiful: a wistful long shot of marching workers across a lake, a soldier putting flowers on his machine gun. There is even some beautiful blood. At the end the dead worker's family are cleaning their windows with beautiful smiles in a beautiful bit of symbolism.

There are some good performances: Peter Schildt as the young lover, Kerstin Tidelius as his mother, Roland Hedlund as his father. There is an enchanting small boy. (Widerberg handles children well.) The girl, Marie de Geer, has a large, sunny face. Her mother is nicely sketched by Anita Björk, more mature of course than she was in *Miss Julie* (1950) but no less attractive. And there are a couple of good scenes: the angry strikers berating the worker-father for helping a wounded strike breaker; a drunken officer, after the slaughter, telling the boss who now scorns him that it was *he* who paid for the bullets.

But much of the rest, often shot in great soft close-ups, is a run-through of the stock symptoms of adolescence and some stock views of workingmen, their games and teasings. When the men are finally at the end of their patience and march in protest, out of nowhere (as far as the film is concerned) comes the *Internationale*. The slaughter is equally unprepared, either in terms of the boss's brutality or the soldiery's oppression. There is not even the sense of a flash accident, something that was not meant to get serious but that blew up. There is simply a sense of omission—a disconnection between most of the picture and its climax.

The obvious comparison is with Monicelli's *The Organizer* (1963), which is about similar struggles in Turin at the turn of the century and which has seemed better to me each of the three times I've seen it. Monicelli's film is rooted in history and in ideological currents, yet replete with the life of individuals: a work that traces aspiration, heroism, futility, and stubborn aspiration on its way to futility again —in other words, it has some sense of the human condition. Widerberg, who has the ability to make nice scenes, has too limited a vision and is too pleased with his prettinesses to come near an achievement like that.

Duet for Cannibals

(November 15, 1969)

Duet for Cannibals was written and directed by Susan Sontag and on that ground, though not much else, demands comment. It tells of an encounter between a young, unmarried Stockholm couple and an older, married couple. The husband is a well-known German Communist activist and teacher, the wife is Italian. The older couple seek to "devour" the younger, not only by utilizing them sexually but by trying to engulf their psyches with a series of tricks. The tricks pile on so thick that it is difficult to know which end is up, in bed or elsewhere, but this confusion seems intended as evocative ambiguity.

The film is a parallel of Miss Sontag's two novels. We are aware that she is versed in the "literature" of the field. We get a string of symbols so blatant that their very blatancy is presumably supposed to make them fresh again: wigs, beards, bandages on eyes, fake deaths, snips of Wagner on the sound track. (This last signifies nineteenth-century romanticism in music, as well as in politics, pursuing the New Left, I daresay.) But, as with Miss Sontag's novels, she simply does not convince us of the need for the work to exist. After we have added up the totals of esthetic apparatus, there is still no affective center in the work. It has no *result* in us (save tedium). I went with some admiration that she had made a picture; I left with some irritation that her literary prestige had got her a chance that people with film talent cannot easily get.

Anyway, to praise Miss Sontag for having made a film at all is really not much more than to praise her for having actually written novels. Films are technically more complicated, but film makers can have a lot of technical help and advice—and refuge. As for the Swedish language, she herself has told us that there was no problem because so many Swedes are bilingual. (She wrote her own English subtitles.) At first I was surprised that, for a critic so concerned with the "visual" approach to film, Miss Sontag shows little that is individual or striking in her ability to see; but then I remembered that her critical essays on the importance of style are themselves written in generally lamentable style.

I liked one sequence. The young couple are in a restaurant, with an obstreperous demented tramp at a nearby table. The manager throws the tramp out, and he lurks outside to harass the young couple in the street. I got a vivid feeling of the random craziness that lies in wait to murder any of us these days, figuratively or otherwise. But these were the only vital moments for me in an otherwise tiresome and arty construct.

Tell Them Willie Boy Is Here

(December 6, 1969)

THIS film, written and directed by Abraham Polonsky, is his second. His first, *Force of Evil,* was made in 1948. For almost twenty years his film career has been in the deep freeze because of political blacklisting. In that time Polonsky has written a novel and (presumably pseudonymous) film and TV scripts; and in that time *Force of Evil* has acquired some fervent admirers. (I saw it only once, when it came

out, and remember liking it.) It was adapted from *Tucker's People*
by Ira Wolfert, a long, grinding novel of some eventual power that
deals with crime and business and their relation. Wolfert collaborated
on the screen adaptation.

Polonsky's chief film work previously had been the original screenplay for *Body and Soul,* directed by Robert Rossen. Both *Body and
Soul* and *Force of Evil* starred John Garfield. The subjects, sport as
brutal business and large-scale urban crime—together with the flavor
of Garfield—typified a phenomenon of the time: the confrontation
of the child of Jewish immigrants with the American metropolis, the
fascination and horror of the power hive and the human gristmill,
combined with the discovery of a vast new arena in which to exercise the perennial Jewish passion for social justice.

The last twenty years have been so tumultuous that one is prepared for an exile to come back completely outdated or strenuously
updated. The surprise in Polonsky's new film is that it is neither one
nor the other but contains modified elements of both.

He has shifted ground visually and symbolically, but his perspectives remain much the same. Like many writers and film makers
with other interests, he has seen in the American West a theater of
metaphoric possibilities. His screenplay, from a book by Harry Law-
ton, is based on the true story of a young Paiute Indian called Willie
Boy who returns to his California reservation in 1909 after some
months of work elsewhere and takes up again with a Paiute girl whom
he loves. They tried to run away once before, but her father stopped
them. (His objection to Willie is unspecified.) Now the father tells
Willie that he will kill him if the boy comes near his daughter. That
night the lovers have a rendezvous, are surprised by the father and
the girl's brothers, and in self-defense Willie kills the father. He and
the girl flee on foot. All this would not have amounted to much in
the ordinary course of things. As Willie says to the girl, "They won't
catch us. They won't even try. Nobody gives a damn what Indians
do." But by coincidence President Taft is touring that part of California at the time, and memories of the assassination of McKinley
raise nervousness about a murderer on the loose in the Presidential
vicinity. (Plus the fact that the local authorities want to look good
while the national spotlight is on the area.) So a posse, including
three Indian policemen, sets out after the lovers. (The sheriff, in a
Joyce-cum-cinema pun, is called Cooper.)

It is immediately clear how such a story, lying in wait in U.S. history as Jack Johnson's story did for Howard Sackler, would spark a man of Polonsky's temperament. It is almost a made-to-order parable on themes of racial prejudice, white guilt transmuted to aggression, Lawrentian blood-life pitted against legally sanctified bloodlust, and the specter of capitalist-militarist society lurking in the ample shadow of President Taft. And the shape of the story—a chase through the California mountains, an arrow-flight from impulsion to death—is quintessentially cinematic.

As for the last—the sheerly filmic qualities of *Willie Boy*—Polonsky has done superbly. He has "seen" his film beforehand, which in itself is not commonplace, and has realized exquisitely what he envisioned. Conrad Hall, who did the outstanding black-and-white photography of *In Cold Blood,* has provided color photography that is above the generally high current American standard; Melvin Shapiro's editing is acute; and with them Polonsky has created a film of fine fabric. I disliked the music of Dave Grusin—a treacly mind trying to be modern and tart—but almost everything that was done after the script and casting were finished is admirable. Polonsky uses some melodramatic visual contrasts—deep night followed by a shot of bright sun so that we actually blink at the glare in the movie theater, but his images of both are so pithy that the melodrama is tempered. Some of his compositions are derivative—the first shot of the isolated railroad station suggests the famous long shot of the homestead in Stevens' *Giant*—but there is a spanking impact in them that makes them valid. Polonsky has used the Panavision screen for its "loud" and "soft" best. There is a wide shot of the posse searching a ford for tracks that in itself says something of nature as refuge and enemy. And there is unusual deftness in the close two-shots and three-shots on the stretchy Panavision screen. One scene between the sheriff and his lady friend, shot under the enormous brim of her fancy hat, embodies the very idea of the close-up: larger but figuratively *smaller,* more intimate and private.

Several of the performances are very good. Robert Blake, the Perry Smith of *In Cold Blood,* is Willie Boy and comes on taut and seamless, ensealed in Willie's experience and expectations. He knows the world he lives in and is ready, his compact body like a small mobile fortress. Robert Redford, the sheriff, achieves here what he missed in *Butch Cassidy and the Sundance Kid,* a performance completely within the film, devoid of references to extrinsic all-American boyish appeal. Susan Clark, as the Eastern young lady doctor in charge of the reservation, has dignity and sexual susceptibility.

But then we come to what we might call (on the basis of *Medium Cool*) the Haskell Wexler touches: the falsities in a film dedicated to anti-falseness. Who can believe Katharine Ross—of *The Graduate* and *Butch Cassidy*—as the Indian girl? Miss Ross plays it adequately, but she looks like a Bryn Mawr girl after a month in Hawaii. Who can believe that this Indian girl, going to a secret rendezvous with her lover and knowing that his life is in danger, would wear a white dress? Just because it looks good in the low-key lighting of the shot? Would she and Willie then have made love in the middle of an open glade by moonlight? Just so that we could see a good composition when the father and brothers surprise them? The line with which the lady doctor recites her credentials is clumsy theatrical exposition, self-consciously trying to get over gawkiness in a medium that affords other ways to handle factual groundwork. When she comes into her hotel room and discovers Redford there, it may surprise *her* but cannot surprise anyone who has seen, say, three films in his life. And at the end, wouldn't Willie Boy, the "natural" man, have heard the sheriff climbing up behind him on the rocks—in those boots? Willie doesn't hear because the finish is needed for the film.

And then we come to what have to be called the Stanley Kramer touches: the assumptions and patterns that translate liberal thought into the knee-jerk responses of liberalistic drama. The characters and issues wear tags, as in a nineteenth-century theater program. At the start the storekeeper and the pool shark are so blatantly anti-Indian that we feel they are being painted into a backdrop against which the story is to be played. (Of course their attitudes are real enough in life, but credibility in art is a different matter.) Similarly, the old-time Indian fighter—played by Barry Sullivan like a wall-to-wall urban account executive—probably *would* long for the days when he could kill at will, but here he has been brought in merely to spout those sentiments, so he is less than believable. The sheriff is given that "human" twist that is now almost as much a cliché as the brutal sheriff: a feeling of communion with the man he is hunting, a sense that the fugitive is intrinsically superior to the posse that is hunting him. We got it from the sheriff in *Lonely Are the Brave* and the sheriff in *The Defiant Ones* (Kramer's best picture, I still think, though I overrated it at the time). I was sorry to see Polonsky dust off that stale novelty.

The two pairs of lovers are contrasted so patently that art shrivels: the two Indians, who want nothing but to live together; the two whites, whose civilized complexities make them want to kill each other in bed. At one point there is even a segue from the white woman's humiliated weeping in her hotel room to the Indian girl's sobbing by her lover in the open air—from the unhappy white society to the unhappiness caused by that society. How can we *not* get the point? What option does Polonsky give us?

He does make some attempt to take the picture out of moral simplicities. There are gnomic utterances: when someone says of the chase, "It doesn't make sense," the reply is, "Maybe that's the sense it makes." The Indian girl's death is deliberately left unexplained. Did she kill herself to unburden Willie's flight? Or did he kill her, in old tribal style, so that she would not fall into enemy hands? These, and other devices, try to give some body of mystery to a morality play; but the bones of the morality stick through.

At the fundamental level of the film's theme, there is no drama of consequence. The characters simply obey the author. If this is how things actually occur in history, that fact is no more artistic justification here than with the bigots' dialogue mentioned above. Nothing really happens to either of the major characters except that eventually one kills the other, which might have happened near the beginning as fittingly as at the end. From the start, Willie is a fatalist, ready to die. From the start, the sheriff is respectful of him (declining to interfere with Willie's illegal drinking on the reservation) and also his congenital enemy—as he still is at the end. The two simply move through the action unaltered and unalterable. It is the lack of *possibility* in them, for change or difference, that reduces them from characters to exponents. There is a substantial difference between predestined figures, pursued by Furies, and puppets in a parable. Across Polonsky's often gorgeous screen moves his invisible blackboard pointer.

All of us have been warned repeatedly that the critical issue of Form versus Content is a corpse, but here the "versus" is so heavily underscored that it puts argument into the corpse once more. On one hand there is Polonsky the filmer, fluent and sensitive, elliptical, with a fine rhythmic sense of when to touch and when to hold, and a compositional sense that plumbs his images. On the other hand there is Polonsky the writer, whose cosmos is bounded by rigid ideas of order without resonance. In the age of Beckett and Pinter, Polonsky's dramaturgy still smacks of early Odets and John Howard Lawson. So for all the excellences of his directing, his film is not very moving. It has the effect of a modern remake of a picture we saw some time ago.

The paradox is that his twenty years of enforced inaction are not apparent in his direction, of which he has done none in that time, only in his writing, at which he has been busy. Still, it is a happiness to welcome him back—if "welcome" is not an offense to a man who has been stupidly and cruelly barred for so long. Other artists have suffered—in their subsequent work—from enforced inactivity, storing up undelivered messages, resolving not to change, as proof that they have not been bowed by adversity. Now that Polonsky is back, we can all hope that he can keep working and that work itself will flex all of his powers equally.

POSTSCRIPT. Later I saw *Body and Soul* and *Force of Evil* again, and both had diminished badly. Both scripts seemed products of a mind which believes that, by pouring on imitation-Odets dialogue, one converts a conventional genre picture and the warped version of a serious novel into works of social significance. Both films have unbelievable endings. As the boxer in *Body and Soul,* Garfield wins a fight although he has accepted $60,000 to throw it. (He doesn't give the money back, by the way, as far as we see.) After the fight, he defies the gambler who paid him, saying, "What can you do? Kill me? Everybody dies." Then he and his girl walk off as if they really believed that this rhetorical gesture would keep him from being killed. The girl, particularly, is made to look phony by this empty Happy Ending.

At the end of *Force of Evil,* Garfield, a crooked young lawyer, goes to look at his brother's dead body on the shore of a river, where he has been told it is lying. (He doesn't even pull it out of the water.) He has just killed the man who told him of the death, as the presumptive murderer, although his brother actually died of a heart attack. Again Garfield is accompanied by that girl friend essential to a big last scene. And his voice on the sound track says, "If a man can live so long and have his whole life come out like rubbish, then something was horribly wrong . . . and I decided to help." The word "rubbish" is meant to have a Gorkian ring, but all that has happened is that his taste for the criminal life was washed away by his brother's accidental death. (If his brother were still alive, he would not consider his own life rubbish.) The transformation is mechanical, and his nobility is suspect because either he would have been caught by the police for the murder, or else the gangster friends of his victim would have killed him.

Force of Evil has further hollowness. Polonsky makes the usual Hollywood pretense of the day, when dealing with Jews, that the characters are not Jewish. Beatrice Pearson is as egregiously miscast here as Katharine Ross in *Willie Boy*. And the dialogue is full of bombast intended to be deep and resonant. Example: "I'll kill you with my own hands rather than let you put the mark of Cain on my brother."

The paradox of *Willie Boy* becomes all the more marked by seeing *Force of Evil* again. Polonsky has been writing since 1948 but his writing has not much improved. He has not been directing, but his direction has acquired a much superior sensibility.

Z

(*December 13, 1969*)

AN exciting film on an agonizing subject. Three years ago Vassili Vassilikos, a Greek writer now living in Paris, published a novel called *Z,* a thinly disguised account of the murder of Gregorios Lambrakis in 1963. Lambrakis, a leftist deputy and a professor of medicine at the University of Athens, had just addressed a meeting in Salonika protesting the deployment of Polaris missiles in Greece when he was knocked down by a small truck; an investigation proved it was not an accident. In Greek the initial Z stands for *zei*—"he lives"— just as, in Italian, *VV* stands for *viva.*

Jorge Semprun, himself a well-known novelist, has written the screenplay of the novel with Costa-Gavras, the director who made *The Sleeping Car Murders*. The film was shot in North Africa, in French, and is shown here with English subtitles. It is flawlessly acted,

sharply edited, and excellently photographed in color by Raoul
Coutard. Costa-Gavras has directed with fire but without much
rhetoric. In the script there are polemical touches; in the direction
there is passion but only a little dice-loading. It is not a drama of
political ideas but of action in the political area. It has no political
thought in it, to speak of—no more, say, than the décor in a John
LeCarré espionage thriller. We see the *results* of political conviction,
in physical and moral courage.

There is no mystery. The film begins with officialdom's intent to
harass opposition—bland harassments while mouthing democratic
platitudes and a pretense that the death was an accident. We know
otherwise. (Neither Greece nor Athens nor Salonika is mentioned, but
shop signs and newspapers are Greek.) The matter would rest there,
sat upon by the beefy rumps of generals and police chiefs, except for
the young investigating magistrate who is brought in to put a quietus
on the matter and who disappoints his superiors. It is he who is the
hero of this film, not the deputy, who is only the occasion.

Thus this is not a story of politics but of a quest for justice, of an
investigator who presumably would have followed the facts wherever
they led him politically. Without such a magistrate, the truth of the
matter would have been irrelevant, and so would the protests of the
dead man's friends. (Most of them were conveniently murdered soon
after.) In fact, the doing of justice didn't really result in anything;
the colonels took over the government in a few years anyway. But
the struggle between idealism and power is always a good subject for
fiction, in film or elsewhere; perhaps one of the justifications of fiction
is to keep that struggle alive, to provide a point of tension against the
world of fact in the newspapers.

For Americans, the added horror of this film is that the muscle behind the Greek government is American. I and many others have
heard Andreas Papandreou, the former opposition leader, talk about
arrant CIA influence on the colonels, which no one has really bothered
to deny. After a short respectful pause, as a brief obeisance to democratic process, American military aid and corporate investment and
tourism resumed. And—a private event of such a character that it is
not immune from comment—the widow of a murdered American
President married a chief financial backer of this government of
unmistakable oppressors and torturers.

Fundamentally, what *Z* dramatizes is something more terrible

than anything we see. The argument for American interference in
Greece is, as usual, that Communism was stopped, although the
recent report of the Council of Europe denies that a Communist takeover was imminent in 1967. If there had been such a takeover, few
of us believe that Communist officials would be less cruel than those
we see here. But, as usual, that argument excludes a middle; and, as
usual, American democracy ends up supporting gross reactionaries
(no matter how labeled) against dogmatic leftists. The murder of the
deputy is a factor of political war; in a noncynical way, we can get
hardened to such acts. But the soul-shaking threat to the future is that
men interested in truth—men like the magistrate, uncommitted to
dogma—may not arise, or may not want to. That may be the real
price we pay for putting iron lids on troubled countries.

The physical impression that this film gives is that it is hurrying
to record certain facts before they are covered over. Motion is of
the essence. Costa-Gavras' camera tracks and dollies almost constantly, yet without dizzying us (unlike a recent Czech fiasco called
Sign of the Virgin), because all the motions are tightly linked to the
impulses of the characters or of the audience. We *insist* that the
camera move as it does, so the tracking both feeds and stimulates our
concern. It looks as if Coutard has used long-focus lenses for much
of his close work, particularly outdoors, which gives many of the
close-ups and two-shots a grainy, unglamorous, almost journalistic
feeling, as if we had been magically allowed to get near. And there
are some shots that are almost salon gems, like one of an official's
head against a white wall with some framed photographs on it—
something out of Erich Salomon.

There are traces of slanting in the script. When the doomed deputy
arrives at the airport, he stops to shake hands with a porter. (Wouldn't
a fascist demagogue do the same?) After he is killed, his wife moons
around his hotel room and weeps over his shaving lotion. (Don't the
widows of murdered fascists miss them?) One of the two murderers is
an aggressive homosexual. (Is that a fascist monopoly in the Middle
East?) All the government men, except for the magistrate, are pompous and slimy. All the opposition men are variously brave and
sincere. But then this film is not a tragedy, at least not in the Hegelian
sense: that is, the opposition of two partial truths, each of which
thinks itself whole. Z is an intelligent drama, intended to whip up
sympathy for one (necessarily) partial truth. Which it does.

Yves Montand as the murdered man and Jean-Louis Trintignant as the magistrate are simple and strong. Georges Geret is appealing as a witness whose testimony is crucial but who is absolutely apolitical and who just wants amiably to report what he saw. Renato Salvatori and Marcel Bozzufi freeze one's marrow as the two murderers. Charles Denner (of *Landru* and *Life Upside Down*) plays a leftist lawyer with his typical immediacy and heat. I have no flat rules about filmgoing, but if I had, one of them would be never to miss a film with Denner.

In the Year of the Pig

(December 13, 1969)

CONCURRENT with reports of the U.S. Army's slaughter of 567 Vietnamese villagers—accompanied of course by President Thieu's quick denial—comes a documentary called *In the Year of the Pig*. A misleading title; the film has nothing to do with Weathermen or Chicago '68. I don't know what the title means.

This is an account of the progressive American involvement in Vietnam, from the beginning, and was made by Emile de Antonio, co-maker of the Joe McCarthy documentary *Point of Order*. De Antonio got his film clips from American TV, from East Germany and Prague, and has interlaced them with good interviews that he made for this picture, with such subjects as Jean Lacouture (author of a fascinating biography of Ho Chi Minh), Senator Thruston Morton, Roger Hilsman, Harrison Salisbury, David Halberstam, Father Daniel Berrigan, and several others. Professor Paul Mus of Yale discusses Vietnamese cultural attitudes, which America disregards. (Two and a half years ago, in a series in the *Atlantic,* Frances FitzGerald made this same point about cultural barriers that force us to view Vietnam through American eyes. The situation seems unchanged.)

Most of De Antonio's picture is straightforward, and therefore heartbreaking. Some of it is facile. (If you juxtapose subsequent events against public remarks, what political figure could not in some way be made ridiculous?) And occasionally the editing is tendentious. (A shot of an old Vietnamese woman crawling among ruins is followed by a shot of a smiling U.S. soldier who may or may not have been looking at her.) But the sum is depressingly forceful.

There are no surprises in it except the scary surprise that much of the story has already lapsed from memory. Remember President Eisenhower welcoming President Diem as a great hero? Remember Madame Nhu? The battle sequences are not as effective as those in Eugene Jones's *A Face of War* because, for one reason, they pay small attention to American suffering in the war and, for another, they are too long—the picture's point is political. In that sphere it makes its mark, showing lucidly how we got into Vietnam, how stupid and corrupt the maneuvers were that got us there, and how millions are suffering every day in a war that virtually no one except the Saigon government wants but that continues simply because people can't agree on a way to stop it.

They Shoot Horses, Don't They?

(*January 17, 1970*)

IT has a lot of the come-on of Harsh Truth-telling, but *They Shoot Horses, Don't They?* is quite empty. The film makers can brag to their mothers or analysts or wives—or whomever they brag to—that their picture is uncompromised and sticks bravely to its "downbeat" ending, but so did Bette Davis' *Marked Woman* (1936). What this film lacks is any information about why it was made—except that it wants to be commended for its forthrightness. Yes, in a broad glib manner, it lampoons "the American way," but that's a pretty big, enveloping tent into which a lot of people crawl without really paying any kind of painful admission. Nothing is easier than free surfing on a wave of national self-excoriation.

Horace McCoy's novel of the same name (1935) deals with a dance marathon in the Hollywood of the Depression years, an event in which couples danced for days in order to win prize money. People paid to watch them struggle, and spectators also offered money for "sprints" when the contestants were particularly tired. The competitors got ten minutes off for rest and food every two hours; the couple that danced 1,500 hours got $1,500.

When the book was published, it had a fair sale, but when it was subsequently published in France, it was a critical and commercial sensation, and Horace McCoy—not very well known here—became an outstanding American writer there. My own impression is that his French success was one more example of old-style inverse anti-Americanism. To praise what was second-rate and brutal about America, like facile gangster films, was a way of patronizing. (It had its bitter consequent reversal in the current hatred among the French young for American brutality.)

McCoy's French acclaim eventually echoed in this country, and his novel began to get overcompensatory critical attention in the U.S. I'm aware that some eminent French writers praised it highly. I don't know what those writers thought of James M. Cain, but my own view, like Edmund Wilson's, is that McCoy is notably inferior to Cain.

Horses is precisely the kind of novel that a certain kind of deep-think movie mentality—say, of the Joseph Strick–*Savage Eye* kind—has always chafed to film and has used as a proof that Hollywood had no guts. It is both tricky and flat, but it nevertheless makes some effect by means of its typographical tricks and the author's consistent flat tone. The film fiddles with the original viewpoint—it is not seen exclusively through the hero's eyes. It fiddles with the revelation of his partner's murder, in attempts to build up suspense. It inserts dubious new stuff and omits important matters. For instance, the pointless death of the girl is prefigured and prepared in the book by the even more senseless death of an old lady fan of hers, Mrs. Layden, who has been watching her from the stands. Mrs. Layden's murder is omitted from the screenplay.

But the film's basic esthetic failure is in its texture—something I've discussed previously in regard to other adaptations. Take this passage from the novel in which Mrs. Layden first appears:

"Hello, Gloria," a voice said.
We looked around. It was an old woman in a front row box seat by the railing . . .
"Hello," Gloria said.
"What was the matter down there?" the old woman asked.
"Nothing," Gloria said. "Just a little argument."
"How do you feel?" the old woman asked.
"All right, I guess," Gloria replied.

"I'm Mrs. Layden," the old woman said. "You're my favorite couple."
"Well, thanks," I said.

That passage—more than anything else in the book—has stuck in my
mind since I first read it some twenty-five years ago. "You're my
favorite couple." Obscene and pathetic, both; and laid in there
nakedly to resound on its own. In the film it becomes a warm little
touch. And instead of the hypernaturalistic black-and-white tone of
the prose, we get the golden glow of color, swirls, lights—all the
apparatus of punctured romance, instead of the bottom of a pit.

The director, Sydney Pollack, has said, in an introduction to the
published screenplay, that when the characters of a novel are put
on a screen, they must be filled in with action and dialogue so that
they won't seem hollow. But the whole point of Gloria is that she
should seem hollow. The filling-in, the provision of little climaxes
and of switches to her viewpoint, make her an aggressive toughie,
almost as if she were conscious of being tough in a movie.

Jane Fonda plays the role, and readers may remember that I have
been enthusiastic about her possibilities as actress ever since her film
debut ten years ago. She has given some very good performances and
some very lazy, absentminded ones since then. Here, in this italicized
character, she supplies a cutting edge that at least gives it presence.
She is made up and coiffured to look like a harpy in a Lynd Ward
woodcut of the thirties, and she gets an effective ham relish out of her
moral slumming. But there is a fundamental flaw, which Pollack has
not caught: Miss Fonda plays this Texas-born tramp with her own
good Eastern finishing-school accent. She says "goddam" as if the
term following were going to be "debutante's ball" instead of "dance
marathon."

Michael Sarrazin, the hero, is too obviously being promoted as a
"sweet" successor to Henry Fonda to be credible as Gloria's partner
in the lower depths. Red Buttons, attended as always by the
stench of theatrical "heart," plays an aging sailor in the marathon
and aids the air of naturalism very little. Susannah York is unbearable
in the equally unbearable role of an English actress who enters the
marathon hoping to attract producers in the stands.

The only interesting character is the one that has been arranged for
Gig Young, the promoter and MC; at least it has more than one note
in it. Young plays it well, with good carny con, and is photographed

not only to exploit his age but to distort his pleasant features clownishly. It's not a hard part, in fact it's a plum; but Young gives it an extra dimension: we feel he is allowing us a peek behind his own mask, the man who wanted to make it as a handsome film star, and didn't, and now is diving into character roles. He almost wallows in the ugliness of the part as he moves, in his own career, from a lifelong concern with looking pretty to the-hell-with-it candor.

There is a lot of good thirties physical detail, but that's no problem if you have a big enough budget for research and re-creation. There's some by-now commonplace subjective use of the camera. What there is *not* is any reason for this film to exist. As a cross-section of American life, it's ridiculous. As a representative phenomenon, it's eccentric and badly dated. As an intrinsically moving story, it's nonexistent. Mc-Coy's novel still has some grittiness, as if one were being dragged by the ankles over that dance floor, face down. But the visual texture of the film and the changes in the original shape transform what was pretty grim into something grimly pretty.

Intimate Lighting

(January 17, 1970)

Intimate Lighting is a Czech film by Ivan Passer, made in 1965, that tries to answer the question: how much can be done interestingly with unexceptional domestic material? The answer is: a good deal— because of Passer's sympathy and skill, the valid performances, and his absolute determination not to get excited.

A cellist and his girl come to stay with a musician friend of his and the latter's family in a small town, to play in a concert there. The two men haven't seen each other in some time, and they swap reminiscences as they view each other's present circumstances. The host's father makes his little lewd cracks about the visitor's girl, his old wife unburdens her secrets to the (charming) girl, and the whole atmosphere is beer and *Eine Kleine Nachtmusik* in the parlor.

There is a bit too much about burping and snoring, and there is an unfortunate sequence in which the girl encounters an actual village idiot—which seems exploitive of the man. But, on the whole, this is a nice relaxing film made by people who understand a milieu, are just sufficiently sentimental about it, and can simply allow it to *be*.

The Circus

(January 24, 1970)

I saw Charlie Chaplin's *The Circus* in 1928, the year it came out, and loved it. I saw it about twenty years later, at a private showing, and loved it. I saw it again recently, in its first commercial release in more than forty years, and loved it again. Obviously the "I" was vastly different each time. One proof of Chaplin's genius is that the same film had the same fundamental effect on three different people, and this is not an egocentric proof because there must be millions who have had similar experiences with his pictures. But with *The Circus,* the experience is particularly striking—even more touching—because, with the last two viewings, I've known that it isn't one of Chaplin's very best.

Naturally, that is strictly a relative statement.

In this film the Tramp is hanging around sideshows, is chased by a cop, and flees—into the main ring of a circus. The audience there, which was bored by the clowns, laughs heartily at this chase. The ringmaster-owner hires the Tramp to make him a clown, but the Tramp can't learn the traditional skits. He is fired, is again chased into the ring—by a donkey—and again convulses the audience. This time the wily ringmaster hires him as a prop man and arranges to have him chased into the ring at every performance. The Tramp is a hit but doesn't know it. The bareback rider, who is the daughter of the ringmaster and whom the Tramp loves, finally opens his eyes to his success. But when he learns that she is in love with someone else, the Tramp loses his ability to be funny. He brings the two lovers together, then lets the circus go on without him.

The Circus is slim, and part of the slimness is purely dimensional— it is Chaplin's shortest feature after *The Kid*. But there is also internal slimness: the story lacks drama and progression, and the supporting characters are not well-developed. The girl—played by Merna Kennedy, the most pallid of his leading ladies—lacks the gamine vitality of the waif in *Modern Times* or the complexity of the dance hall girl in *The Gold Rush.* (I saw the latter again recently and noted once more how subtly Chaplin assumes that we know she is a prostitute, without ever saying so, and proceeds to build the Tramp's pathos on the substructure of that assumption.) The only mildly colorful character here is the ringmaster, the girl's father, who in some books is called her stepfather. His brutality to the girl is stock villainy, but his handling of the circus has cynical verity.

Still, if *The Circus* lacks drama in its story, the dramatic structure of its comedy episodes is marvelous; and if the characters are thin, the thematic implications are not.

As for the first, Chaplin only occasionally uses an isolated gag. For instance, when he is dusting things in the circus one day, he takes the magician's goldfish out of the bowl, wipes them off, and puts them back. But, for the most part, the comic incidents are part of a knitted structure that grows to considerable size from a quite small beginning. I'll trace one example.

A pickpocket lifts a fat wallet and a watch from an old man in the sideshow crowd. The old man discovers the theft, and the pickpocket quickly shoves the stolen goods, unobserved, into the pocket of the Tramp, whose back is toward him. The starving Tramp is now (1) an innocent thief and (2) the unwitting possessor of the means to end his hunger.

That is the simple beginning; now see what happens. In front of a hot dog stand, the Tramp steals bites of a hot dog held by a baby over his father's shoulder. This would be funny anyway, but we know that the Tramp has enough money in his pocket to buy the whole hot dog stand. While he is gazing longingly at the franks, the pickpocket approaches and tries to lift back the loot. This time a cop catches him with his hand in the Tramp's pocket and restores the loot to the "rightful" owner.

Amazed at his luck, the Tramp orders about a dozen franks. While he waits, he pulls out "his" watch grandly and looks at the time—just as the old man passes whose watch it really is. When the old man sees the Tramp also pull out his wallet, he calls a cop. The Tramp runs. Meanwhile the pickpocket has escaped from *his* cop. In a wonderful tracking shot the Tramp and the pickpocket run toward us side by

side at full speed, each one with a cop behind him. Then the Tramp
raises his hat to the pickpocket in salute, and they split in different
directions.

But they meet again in a funhouse to which they flee, where
Chaplin hilariously mimics a life-size mechanical doll and where there
are marvelous optical tricks in a mirror maze. (Did Orson Welles
remember this scene when he made *The Lady from Shanghai?*) At
last the cop chases him into the circus ring and, after further complications with a magician's apparatus, into a new life. The Tramp escapes,
returns the wallet and watch *en passant,* then hides in a chariot and
sleeps. Next day he is discovered and hired.

All this exploded out of a tiny capsule: the pickpocket putting his
loot in the Tramp's pocket. After that, all that was needed was the
insanity of strict logic. But it is not "mere" logic, it is a dramatist's,
including by implication everything that has gone before. The progress
is not linear but cellular.

The theme of the film is a contrast between life and art. The Tramp
is funny as a man. He is unfunny when he tries to amuse in a circus
skit. His funniness is in his spontaneous being; consciousness kills it.
There is one brief period during which he knows he is being funny,
after the girl tells him he has been a hit for some time. With that
confidence he can continue; but he loses that confidence as soon as he
learns she doesn't love him, and he cannot "make" funny again.

Chaplin was always fascinated with this confrontation of a funny
man with a theatrical environment in which people are trying to make
emotion. In one of his earliest shorts, *A Film Johnnie* (1914), he
follows some actors into a studio and bursts open several contrived
scenes with the reality of his being. In one of his last features, *Lime-
light* (1952), he is still concentrating on the dividing line between
facing audiences with comic equipment and facing life with comic
bravery. At the end of *The Circus,* in one of the loveliest shots in any
Chaplin film, he stands on the circus grounds as the wagons roll away
one by one. Finally alone, he sits, and we see he is in the middle of a
circle that was marked on the ground by the circus rings. The theater
has once again become the world.

Chaplin has added a score for this new release, which is adequate
though full of sudden cutoffs, and he includes a song, sung by himself
at the beginning—"Swing, Little Girl"—as Merna Kennedy dangles
idly in a trapeze. In best kindness, let us say that the song is an echo of

a bygone age. But then so is the courtliness with which he treats the girl he loves or the pickpocket from whom he's parting. So complete is the magic with which this chivalry envelops him that it gives him a physical magic: his wing collar is always neat and clean whether he is in a circus, an alley, the Alaskan gold rush, or sleeping by a roadside. To my knowledge, no one has ever objected to the impossibility of this haberdashery—nor do I: which is one more proof of the persuasion of genius. We *want* Chaplin to be as he is for our sakes, ducking with polite aplomb and clean collar the outstretched hand of the monster policeman of the universe. Hart Crane understood this. When he wrote "Chaplinesque," he spoke of Chaplin as "we":

We will sidestep, and to the final smirk
Dally the doom of that inevitable thumb
That slowly chafes its puckered index toward us,
Facing the dull squint with what innocence
And what surprise!

A Married Couple

(January 31, 1970)

RECENTLY I heard a remark made by the late Sidney Meyers, a gifted film maker: "Artists were too happy, so God invented film." God got his wish. Now look at the troubles, not only for artists but for audiences. With every advance or benefit, there is at least one new problem. Look, for instance, at the nonfiction film. What a blessing it seems to be. What troublesome questions it raises.

There are two principal kinds of nonfiction film: first, the kind in which the subjects are busily concentrating on something else—fighting a battle or running a steel mill or pole vaulting—or in which they are unaware of being photographed because the shots are "stolen"; second, the kind in which the subjects are well aware of the camera's presence, in varying degrees of consciousness—the kind now called *cinéma vérité* or direct cinema or actuality drama.

Allan King, the Canadian who directed the unforgettable *Warrendale,* about emotionally disturbed children, makes this latter kind.

Here are the first two paragraphs of the publicity release on his latest picture:

A Married Couple was filmed on location: the major part of it in the Toronto home of its stars, 42-year-old Canadian advertising man Billy Edwards and his American wife, Antoinette, 30, their three-year-old son, Bogart, and a dog named Merton. Cameraman Richard Leiterman and soundman Christian Wangler spent ten weeks in the Edwards' house, often from six o'clock in the morning until long after midnight, rarely talking to them, never eating with them nor becoming part of the family, but always observing, filming and recording their daily life. The interior of the house had been entirely lit before shooting began, so the only other members of the film crew were director Allan King and an assistant cameraman.

Immediately interesting. But, almost as immediately, a problem. Usual critical values (whatever one's may be) for fiction films or for "nonconscious" documentaries are secondary with *A Married Couple*. Primary are questions of psychology, society, image, and reality. The key word in the paragraphs quoted above is "stars." Much has already been written about changes in the concepts of privacy that *cinéma vérité* brings about. But now it's becoming clear that *cinéma vérité* is, to some degree, only recognizing changes that have already taken place. Several months ago I saw a TV interview with a sweepstakes winner, a lady who had spent her life tending her corner grocery for thirty years, with no hint of publicity or spotlight. Then her ticket won and the TV crew arrived and the reporter asked, "Mrs. Brown, how did it feel to win the sweepstakes?" She smiled and began in best celebrity-at-ease style, "Well, Ted . . ."

Well, Ted . . .! Like Sammy Davis, Jr. being Serious on the Johnny Carson Show. Mrs. Brown, citizen of a world soaked in film-TV consciousness, a world in which images of self are more a part of self than ever before in history, was instantly ready, even after thirty years in the shadows. How can the cameras of King and others *intrude* on the lives of people who are now waiting, not for Godot but for Godard?

I met Allan King in New York while he was making this film. He spoke of it only briefly, saying that it was about a couple who might or might not break up, he didn't know. I didn't ask—but couldn't help wondering—which he was hoping for; which the couple were hoping for; what effect the camera would have on their decision, how they would decide to *play* it. And yet, of course, play it for real.

The meat of this film is daily life: the baby, the car, the job, the cycles of small-quarrel-and-reconciliation, the larger differences that are left unsolved on screen. Seventy hours of film were edited down to ninety-seven minutes, but even so there are longueurs. Unlike the subject of *Warrendale,* most of married life is not dramatic. If King had cut only to the most dramatic moments, he would have falsified his rendition of a marriage. Some tedium is of the essence of his method, and it helps the sense of eavesdropping.

But that sense is by no means consistent throughout. Part of the time I forgot I was watching people who knew there was a camera present; the reality became the same as in a credible fiction film. But part of the time I was aware that they were camera-conscious, either because they glanced at it or because their actions and language took on a touch of bravado, as when the language became salty in an almost studied way. (This also happened in *Salesman,* by the Maysles brothers.)

There are moments in filmgoing when what I am chiefly curious about is the process of filming. With most skin flicks, I would much rather spend five minutes in the studio watching the amatory couple get ready for a take, with electricians and grips and cameramen bustling about, than see the final take in CinemaScope. There were similar moments in this film—for instance, a sequence in which the couple are in bed together at night and he teases her into having sex with him. The teasing begins to work. End of sequence. I wanted to know what had happened there, before and after. How did they get to feel free enough to do this at all? I wanted to see the crew and King, as much as the couple. What happened when she began to yield? Did they wait for King and the crew to depart? Or not?

Much of the time there seems to be a threefold level of consciousness in the Edwards couple: awareness of the camera, sporadic forgetting of its presence, and then an *adjusted* awareness—the stages one goes through with, say, the hum of an air-conditioner. This threefold awareness is particularly striking in such a sequence as the one where they talk about their desire to be famous—this obscure couple sitting in their modest Toronto house, wishing they could be famous, at the very moment they are starring in a Technicolor movie. But some of the best sequences seem to me to be singlefold, when camera-consciousness at any level is imperceptible. (This, too, was true of *Salesman.*) There is a quiet scene in which their big old dog is lying on their bed. Edwards stands next to the bed for a moment, looking at him, then throws himself down face to face with the dog and simply stares at him. A bit of silence; then cut.

Here, as in *Warrendale,* King's direction consisted only of his choice of material, his personal reticence, and his editing—which leaves out a great deal that is usually subsumed under direction. It is on the first count that he showed his most acute cinematic instinct— not so much by his faith in the interest of the Edwardses' lives as, quite specifically, the interest of their faces. At first view, neither of those faces is particularly appealing, and I felt, "Do we have to watch *them* for an hour and a half?" Through the course of the film, her face became witty, lovely, quite homely, silly, and sad, a much greater range than it promised. His rather unassembled face, made up of features that don't seem really to go together, took on an organic entity; the camera assembled it *for* him, over a period of time. This is by no means automatically true of most faces, which often seem to recede under prolonged scrutiny. King's faith in the way these two faces would bloom, cinematically, is his subtlest achievement here.

Is it a good film? For me, the only answer is that some of it satisfied my instinct for snooping, some of it dragged, some of it seemed tacitly conspiratorial between the Edwards couple and the camera. But in the circumstances it is surely pertinent to ask whether it was a good film for this couple. They overcame their marital difficulties— pro tem, anyway—and now have a second child. They are said to be lonely without the presence of the camera and the lights (which were strong enough for color photography). If so, they are only being honest about what is implicit in our society. Nowadays we're all on camera, Ted.

M*A*S*H

(January 31, 1970)

WE'VE had black humor, and will continue to have it; but now we also get red humor. Laughs—and I mean side-splitting laughs—in the middle of spurting blood. *M*A*S*H* is the Army abbreviation for Mobile Army Surgical Hospital. (The inserted asterisks are Filmland's own Hyman Kaplan touch.) This comedy takes place in a military hospital three miles behind the lines of the war. Which war? You remember—the one we dropped in there between World War II and Vietnam. What was it again? Korea. Right. Some 33,000 American battle deaths, but it slides past.

The script was written by Ring Lardner, Jr., from a novel by Richard Hooker, reportedly a pseudonym for a well-known surgeon. The tenor is familiar service comedy about grifters having a good time amidst the gore, except that these surgeons and orderlies and nurses are *really* amidst the gore, which we constantly see, and are joking and making dates while they remove shrapnel from hearts or saw blithely away on the useless legs of patients who are always just out of sight. And why is it funny? Because the doctors are sane and are remaining sane; they are not laughing to keep back the tears. First, they are surgeons (and we all remember what Freud says about them); second, they are professionals; third, they are powerless to alter their world; fourth, even if they can't alter it, they still believe in the life on which their world is spitting, and they spit right back by saving life, if they can, lightheartedly.

Some of the gags are soggy and protracted and stale—sort of *Catch–2,222.* The first few minutes almost drove me out of the theater. A later football game between two medical units is TV comedy at its hokiest (except for the language). But much of the rest, even the usual sex jokes with nurses, is unusually funny. The language is rough and true, the humor is quick, the acting is good revue-sketch Ping-Pong, and the editing is generally ahead of us by the requisite shade of a second.

Elliott Gould, of *Bob & Carol & Ted & Alice,* is a heart surgeon here—possibly the most amusing heart surgeon ever. Donald Sutherland, who was an English lord in the abominable *Joanna,* is a likeable shuffling, ugly fellow-doc. Roger Bowen is winningly impotent as the CO. A girl named Jo Ann Pflug plays Lieutenant Dish, and rightly so. Robert Altman, who has directed a lot of television and a few films that I have not seen, has handled most of the action with fine impertinence and brisk pacing.

It could be Vietnam as well as Korea, and of course it really is Vietnam as well. If there are parents and sisters and wives who object to jokes about the atmosphere in which their men are being treated, these doctors seem to say, "Do something about it, then. As long as you keep on being the silent majority, we're going to keep on whistling while we work."

The Milky Way

(*February 7, 1970*)

NOT much in art is more pleasant than the sight of an old artist continuing to grow. Most old artists—even great ones—repeat, or lose their creative vigor, or maunder, or quit. But some keep growing almost as if their previous careers had led purposefully to their last years—for instance, Verdi, Ibsen, and Yeats. Equality of stature aside, this is what seems to be happening with Luis Buñuel. He is now seventy and has announced his retirement more than once but has kept working; and of his last three films, all made since 1965, two are the best Buñuel that I know. The exception is *Belle de Jour,* which, though it was sumptuous, showed his old weakness for aggrandized sensation masquerading as meaning. But the forty-two-minute film made the year before that, *Simon of the Desert,* is a masterly little work of religious irony and religious spirit; and now comes another religious film, *The Milky Way,* which is even better. It is another parable of Christianity, but it is free of Christ parallels (*Nazarin*), of sterile and protracted allegory (*The Exterminating Angel*), of shallow Evil-as-travail-toward-Good (*El*). The structure of this new film is taut and well-modeled, the interplay between idea and image is delightful, the whole work is funny and bitter and peculiarly devout from beginning to end.

In essence, *The Milky Way* is a symbolic history of the Roman Catholic Church in Europe, told in terms of its progress through heresies and schisms. (A postscript guarantees that every theological reference in the film is exact.) The title itself is a kind of celestial pun: the constellation that we call the Milky Way is also known in Europe, it seems, as the Road of St. James. Two present-day tramps, a young man and an old, are taking that road on earth. They start from Paris on foot to visit the tomb of the Apostle James in Spain at Santiago de Compostela (St. James of the Field of Stars). They are just tramps, not particularly religious, and their choice of destination seems more happenstance than zeal. En route they meet a number of odd people and are involved in a lot of strange incidents, some of which take place in earlier historical periods and all of which they accept without amazement, simply with varying degrees of interest. Each of these encounters dramatizes an important heresy or theological dispute in Church history, and almost all of them are handled wittily. One instance out of many: the tramps see a fierce duel between two eighteenth-century gentlemen, but it is not an affair of honor or the heart; one duelist is a Jansenist, the other a Jesuit, and as they lunge and parry, they exchange arguments about the nature of divine grace.

Interwoven with the story of the tramps' pilgrimage are Gospel episodes, with Jesus and Mary and the Apostles. The tramps never meet the Holy Family, but there is a near miss. At the end, on the outskirts of Santiago, the two men encounter a prostitute (who was prophesied to them), and she tells them that the town is empty of pilgrims because the body in the crypt has been proved to be the heretic Priscillian, not St. James. So instead of proceeding into town, they take the woman into the woods to frolic with her. The camera pans from them to two blind men moving through the same woods who soon meet Jesus and his disciples. Jesus restores their sight "according to their faith." The point is obvious. The tramps are merely amiable earthly travelers and "see" nothing; the two blind men have faith and "see" Jesus, who gives them back their earthly eyes. But the point is made painlessly.

The film is as well made as Buñuel's pictures have often—not always—been. Occasionally he still lets people walk out of shots, leaving us to stare for a second at empty places where the action *was;* and occasionally there is a meaningless emphasis (like a close shot of the wheels of a railroad train arriving in Tours). But for the most part there is discretion, the sense of a mordant eye, and the overall feeling of flow that we get even in lesser films like *Viridiana* and *The Diary of a Chambermaid.* The color photography of Christian Matras is lovely, and Buñuel has made special use of it in the costuming of the Gospel scenes. Jesus and the others in those episodes are all dressed in solid but soft colors, suggesting what the colors in Piero della Francesca's frescoes at Arezzo might have been before they faded, which contrast with the complicated chromatic patterns of the latter-day clothes. Buñuel has not forgotten to include some of his hallmarks: a dwarf; a close-up of metal penetrating flesh—here we get spikes through the hands of a nun begging to be crucified, instead of the customary knife. Even these matters, though, are handled with unusual restraint.

But the superiority of this film, like that of *Simon,* is not in the filming as such but in the script. (Incidentally, the cast of *Simon* includes two palmers— pilgrims—from Santiago de Compostela, so the subject has evidently been haunting Buñuel's mind.) It is Buñuel the author, more than the *auteur,* who has made *The Milky Way* the fine work that it is. He has written or collaborated on the scripts of most of his films, and he has worked in three languages—Spanish, French, and English. This picture is in French, and he had the assistance of Jean-Claude Carrière, who collaborated on some of his previous films made in France. After one has enumerated the various elements in the script of wit, slyness, compassion, and human bewilderment, there is left the central and controlling vision; and the source of that vision is to me the most interesting and revealing aspect of the film. It is the artistic concept with which Buñuel began his film career, from which he has often divagated, and to which he has recently returned to use with a new depth and power—the concept of Surrealism.

Some years ago Buñuel said:

It was Surrealism which showed me that life has a moral direction which man cannot but follow. For the first time I understood that man was not free. I already believed in the total liberty of man, but in Surrealism I found a discipline to follow. It was a great lesson in my life. It was also a great step forward into the marvelous and the poetic.

That dialectic between liberty and discipline, resulting in a synthesis that is "marvelous and poetic," has an analogue for Buñuel, I think in the dialectic between God and Church, the synthesis of which is in *Simon* and *The Milky Way.* His vision of religion is a Surrealist one.

The term "Surrealism" was first used by Guillaume Apollinaire in 1917, but the famous definition was made by André Breton in 1924.

In the first of several manifestoes on the subject, Breton said: "I believe in the future resolution of the states of dream and reality, in appearance so contradictory, in a sort of absolute reality, or *surréalité*, if I may so call it." Buñuel was early associated with the Surrealist group in Paris. Salvador Dali collaborated on his first two films, *Un Chien Andalou* and *L'Age d'Or*. Buñuel's photograph is in a Max Ernst collage of members of the group (1930), and the cool razor-eyeball sadism of Buñuel's first film is antedated by the cool finger-piercing of Ernst's painting *Oedipus Rex,* done in 1922. (Ernst himself is in *L'Age d'Or.*) Today Buñuel's Surrealist view rests on a more sophisticated and mature base than the juxtaposition of incongruous objects and acts. Dead donkeys on pianos or cows on beds no longer serve as adequate manifests of dream-reality. True, *The Milky Way* scrambles historical periods, refuses to explain how a priest can be inside and outside a bedroom at the same time, and has sweet little schoolgirls fulminating anathema, but these concretized fantasies do not exist merely to shock and stimulate and invigorate our protocols of vision. Underlying them is a *philosophic* vision that is itself Surrealist: a vision that lifts the history of the Church off the ground into one grand condensation and sustains it through an idea of faith that is larger than any pettifogging theological pedantry. Buñuel is no longer interested in extending consciousness through a series of visual puns and oxymorons, however cruelly scintillating, for which the scenario seems almost an afterthought and justification. With *The Milky Way* the process seems reversed. In his old age Buñuel has come back to Surrealism with greater purity, with a deeper perception of liberty through discipline. He has begun this picture with a very clear, almost programmatic dream reality, a surreal *conviction,* and the visual metaphors proceed firmly from it.

As the two tramps, Laurent Terzieff and Paul Frankeur are amiable and easy. Like many directors outside the United States, Buñuel has been able to use prominent and accomplished actors in quite small parts, which gives his film a fine-grained detail that is often lacking in American films. (And for those who know those actors' faces, it also provides another level of intra-cinema pleasure.) Some of the brief appearances: Michel Piccoli as the Marquis de Sade, Alain Cuny as a mysterious prophet, Pierre Clementi as the Devil, Claudio Brook (who played the title role in *Simon*) as a bishop,

Julien Guiomar as a country priest, and Delphine Seyrig as the prostitute.

I saw *The Milky Way* at a screening that was largely attended by Catholic religious, and I happened to sit next to two priests. When the film finished, Priest One turned to Priest Two and said, "Mad." Priest Two nodded and said, "Mad." I think that Buñuel, a lifelong heretic-devotee, would have been tickled.

Patton

(March 7, 1970)

THE astonishing achievement in *Patton* is that of the producers, Frank McCarthy and Franklin J. Schaffner. Not so much for anything in the picture, although there are some very good things in it, but for the foresight to make it and for being able to coax up the money when they did. A large-scale film like this one, with many locations and tons of equipment and hundreds of complicated effects and hundreds of people to transport, has to be planned at least three years before the public gets to see it. This means that the decision to produce *Patton* was made no later than early 1967. Lyndon Johnson was President; sentiment against the Vietnam war seemed on the rise; the press was full of protest. Sentimental patriotics and adulation of the military were not exactly filling the air. Yet that was when McCarthy and Schaffner began. Through the intervening thirty-six months, the production has traveled like a space missile seemingly headed nowhere that gradually zeroes in more and more accurately on its target—and now *Patton* has landed right on the bull's-eye: the age of Nixnew. The time of the silent majority.

Fortuitous or not, it is wonderful timing. I can't remember anything as opportune since the publication of Cameron Hawley's novel *Executive Suite*—about the virtues of big business—just after Eisenhower's first election. For a lot of years, that silent majority has been hungry for pictures like *Patton,* with a hunger that a picture like *The Green Berets* (though it made money) could not really satisfy. Here is a film that is made carefully, photographed superbly, and directed generally well, with an irresistible performance in the leading role, marvelous battle effects and—above all—an air of intelligent candor. It seems to say, "All right, now, we've had enough of this bellyaching. War is *in* us and we might as well face it. The urge to kill—hell, the enjoyment of it—is in us, so let's not kid ourselves. And—at the risk of sounding corny—what's so damned wrong with a lump in the throat at the sight of the American flag?" Perfect. I saw *Patton* in a large theater with a large audience. The very first shot is an American flag in vivid color filling the wide, wide screen. Some defiant applause. Then out steps General Patton, minute against the immense banner, and I felt the audience lunge toward him with relief. Everything was all right again, the old values were safe. Before Patton had finished his address to his new soldiers (which is the prologue to the film), profane, soldierly, paternal, tough, before the picture had really begun, it was a solid hit.

McCarthy and Schaffner had either prescience or blind luck. But, keeping an anchor to windward, they subtitled their film "A Salute to a Rebel." (Patton's rebellion consisted of being more militaristic than the military!) They commissioned a screenplay by Francis Ford Coppola and Edmund H. North that emphasizes the contradictions in their hero. Patton is well read in military history and has a sense of the past, he believes that his soul has transmigrated through the great wars of history, he writes and quotes poetry, he speaks French fluently, and he is a gourmet. (The only "humane" touch omitted is the fact that he vomited when he saw a German concentration camp.) Also included is the Patton who wished he could give medals to the daring German pilots who had just bombed a lot of his men to pieces, who confesses that he loves visiting a battlefield where men lie dead, who slaps a soldier hospitalized with battle fatigue (thus maiming his own career), and who shoots two donkeys that are a peasant's property because they are holding up a military advance (the incident used in Hersey's *A Bell For Adano*). All these contradictions present Patton as a complex man, which he undoubtedly was. The picture asks our praise for that fact, for not presenting him as a monochrome comic-book swashbuckler. What it omits is that most men are complex, and that some complex men are more admirable than others.

The chronicle follows Patton's ups and downs from the Kasserine

Pass in 1943 to the end of the European war in 1945. Franklin
Schaffner also directed the film, and has done it surprisingly well.
There was little in *The Best Man* and much less in *Planet of the Apes*
to suggest the sureness and the comprehension of spectacle that he
shows here. *Patton* is not remotely epical in the sense of Jancsó's
The Red and the White: there is no grasp of the currents of history,
only a compilation of spectacular events in chronological order; but
Schaffner handles them vigorously. He repeats some cinematic ideas:
over and over he begins a grim sequence with pretty objects: a shot
through a field of waving flowers before we move up to an ambulance,
a shot of snow-covered trees before we pan to soldiers advancing
across a white field. Occasionally there is a long shot at a moment
when Patton is saying something important, and the importance is
diminished even though the words are heard. But for the most part
Schaffner has kept his film vigorous, varied, and tight. And he has
provided an implicit commentary on the juxtaposition of two cultures
by getting the most out of the locales through which Patton moves—
North African buildings, a Sicilian chapel, a Corsican villa, a French
château, a London apartment, among others. All these tell us something about the continent that is being fought over, in contrast with
the man who is leading part of the fighting.

For this, and more, Schaffner is indebted to his cinematographer,
Fred Koenekamp, who shot the film in a new wide-screen process
called Dimension 150, which articulates every plane and reproduces
every color faithfully. The score by Jerry Goldsmith is banal.

George C. Scott, as Patton, is truly commanding, fulfilling the
hard, manic streak that was apparent in him years ago when he was
playing Richard III and Jaques and Shylock for Joseph Papp in
New York. Scott was never shy of self-confidence and now he has
brought up additional power to support it. His voice has hoarsened
rather than mellowed through his career, but, in the latency of explosion with which an actor makes us pleasantly nervous, he is the
most remarkable star since Brando. Karl Malden is, once again, uninteresting as General Bradley. (Bradley himself was military adviser
on the film.) All the British and German commanders except Rom-
mel are portrayed as noodles.

The violence in this film is a very different matter from the violence
in, say, *Bonnie and Clyde* and *The Wild Bunch*. *Patton* is about a
man whose profession is to get other people to act as vicars for his

violent urges, and it makes a clear point that it presumably did not intend. For generals, peace is hell.

Zabriskie Point

(March 14, 1970)

MICHELANGELO Antonioni, one of the finest artists in film history, has made a mistake. I want to hazard some guesses about why he made it, to investigate the relevance of this mistake to current film and to art generally, all of which seem more important to me than his new film itself, but first some comment on *Zabriskie Point.*

As everyone must know by now, it is set in the United States. Mark, a young radical, meets Daria in the Southern California desert. He is fleeing the Los Angeles police, who think that he shot a policeman at a college protest. She, although temperamentally a swinger, is working at the moment as a secretary in Los Angeles and is driving to Phoenix, where her boss is having a business conference in his desert home. Mark, who has stolen a plane in L.A., runs out of gas after he has teasingly buzzed Daria's car. She arrives where he has landed and agrees to give him a lift to get more gas. En route they reach Zabriskie Point in Death Valley, the site of ancient lake-beds, now a frozen heaving sea of borates and gypsum. There, after some talk, they make love.

Later that day, he flies back to return the plane in Los Angeles. (When she says it is risky, he says, "I *want* to take risks.") The waiting police kill the presumed cop killer, who actually is innocent, when he doesn't stop taxiing promptly. Daria hears the news on her car radio. She arrives at her boss's house but very soon leaves. She stops and looks back at the luxurious house, and in her mind she sees it explode, over and over again, its contents floating dreamily in space. Then she drives out of the last shot, leaving us to look at the sun in the west.

The script is the work of the director and Tonino Guerra, who has collaborated on all of Antonioni's best films. Here, instead of the as- sistance they have sometimes had from other Italian writers, they had the help of the young American playwright Sam Shepard, the American radical journalist Fred Gardner, and Clare Peploe, an English young lady who was a production assistant on *Blow-Up*. The fundamental trouble with the film is the script, and the trouble with the script is that it does indeed seem the collaboration of all these people. *Blow-Up* was written by Antonioni and Guerra (one germinal device was taken from a story by Julio Cortázar), and then was put into English by the young English playwright Edward Bond, but it was essentially an Antonioni-Guerra product. That does not seem to have happened here. The script seems the outcome of a poll, taken among all the collaborators, as to what would represent youthful dissent and revolution in America today. The result is like a checklist, accurate enough, but it could be almost anyone's checklist. And the story rests on tenets of revolution which may or may not be valid but which here are rather grossly assumed, not dramatized.

Antonioni's critics have called him a tourist, yet this film does not seem to me even a good tourist's notebook. These are not *his* "notes," they do not have the mark of his personality, as his "notes" about London had in *Blow-Up*. What is primarily missing here is not conviction of idea—he may well be convinced about the force of radical youth and the need for their revolution—but confidence as artist. He is like an actor playing a role in a foreign language who has learned his lines phonetically.

In *Blow-Up* the result was quite different, even though he knew less English then than he knows now, because the picture was an expression of a life view that was centrally his. The London photographer was involved in a mystery of existence that also involved Antonioni. There is no equivalent feeling in the new film. Nothing in these two characters resonates anything that we feel to be truly his; and no "montage" of their characters creates a third force that represents him.

Constantly, we feel him trying to connect with the film, to supply "characteristic" touches, with which he can lay claim to the work. The uncertainty about whether Mark really shot the policeman at the beginning is Antonioni's enforced ambiguity, made mechanically by cutting away; later, Mark tells Daria plainly, "I never got off a shot." The disappearance of Anna in *L'Avventura* was *her* disappearance, not anyone's device. This business of the policeman is a strained

effort to lend ambiguity to a bald story. Daria visits a desert hamlet
—en route to Phoenix—to see a friend who has brought some under-
privileged L.A. children out there for rehabilitation. She never finds
her friend, but she does meet the children, who change from little
scamps to little demons. But the change seems manufactured, the
supposedly haunting air of evil seems to have been blown in by a
wind machine. When Mark and Daria make love at Zabriskie Point,
their intercourse is "amplified" by the sudden appearance of groups
of other young people making various kinds of love until there is a
landscape of passion imposed on the dead terrain; but it is imposed.
Antonioni has said that the fantasy is in Daria's mind, making love
just after having smoked pot, but there is no ground for this in what
we see (in contrast to her explosion fantasy at the end). It seems
merely a strained effort at imagination.

There are beautiful moments. Not the shots of the desert, well done
though they are, because the desert is a subject that a cinematographer like Alfio Contini could hardly fumble. I mean such moments
as the shot from a helicopter that spirals down next to the grounded
plane in which the dead Mark is lying, surrounded by police cars;
it is like the descent to earth of a snared spirit. There is a shot of
Daria's boss in his high office, with a huge flag billowing outside his
window and another skyscraper beyond it, so good in itself that it
goes past obvious satire to a poignant statement of slick, barren
"office" civilization. A highway patrolman stops Daria in the desert
(he thinks she is alone; Mark is hiding), and after she has made an
impertinent reply to a routine question of his, he looks all around the
huge solitude slowly, presumably toying with the idea of ravishing this
luscious, irritating piece—then gets back in his car. His long pause
is one of the few truly implicative moments in the film. When Daria
imagines the explosions of her boss's house at the end, the first of
those explosions is silent—an excellent touch which adds to the
shock and which acts as the necessary bridge into her fantasy. And
the shots of things floating slowly—materials, not people, because
the enemy is things—are nicely sardonic.

But scenes that are supposed to be more directly sardonic, more
sharply critical of materialism—business conferences, TV real estate
commercials with mannequins that live All-American lives—are
so blatant that they boomerang. They do not expose the hollowness
of the culture but the insecurity of the artist who wants to expose

the culture. His disgust is superficial, and his means of expressing it are trite. Compare these business sequences with the stock exchange sequences in *Eclipse,* which I saw again recently and which grows in stature. Those hectic stock-trading scenes are distillations of the culture in which they occur, not self-limited, not cartoons but the essence of power and sex drives as they pass through a revealing flame, grotesque ballets danced against impotence. We see humanity revealed there by the dehumanizing process itself. But the business scenes in *Zabriskie Point* seem the work of a sophomore who has never been near such a meeting; they have no touch of the human reality that makes their banality and their pompously disguised greed so terrible.

Antonioni's insecurity extends to his selection of the two leading performers. He has chosen two unknowns, two nonactors, Daria Halprin and Mark Frechette, neither of whom grows during our two-hour acquaintance. The dialogue assigned them is both sparse and generally tinny, and neither of them has the personality, let alone the talent, to fill out the gaps of silence or the flatness of locution. Antonioni has done some bad casting in his career—for instance, Richard Harris in *Red Desert,* Betsy Blair in *Il Grido*—but this is the first time he has used completely inexperienced people in leading roles. Here it is not so much that he has cast badly; it is more that he seems to have lost faith in the processes of art, has relied on the fact of youth to supply the truth of youth. This use of nonactors, as the best of neorealism has shown, can sometimes work, but hardly when the director is a stranger in a strange land, literally and figuratively.

From the choice of Death Valley as a symbol of American civilization, to the inclusion of gag signs on barroom walls to the shots of garish billboards, this film sticks to the surface, stranded.

How did it happen? When an artist of Antonioni's magnitude makes a mistake like this, more must be done than merely to note it. *Zabriskie Point* is not just lesser than his best work, as *Blow-Up* is lesser than *L'Avventura;* it is hardly recognizable as his work, or as a development from it. Those of us who have loved his art cannot help wondering what happened.

Writing about *Blow-Up* three years ago, I surmised that the defects of that (far superior) film were attributable to the fact that Antonioni was a stranger in London, and I expressed concern that this highly societal artist had just announced plans to make his next film in another strange society, the United States. But even in London he was more at ease than in America. Britain is still Europe, whatever the British say, and if Antonioni was in a foreign country (where in fact he had worked before, to make a part of *I Vinti*), it was one with some affinities for any European and some direct cultural connections. ("Linking our England to his Italy," said Browning.) In England, Antonioni was able to translate Antonioni into English. There is no such translation in *Zabriskie Point,* even though he is here trying to speak like an American, trying to comment from within. I know that he has denied this intent, that he has claimed only to be an observer, but the film stands apart and contradicts him: it purports to be speaking knowledgeably from the center of American social turmoil. If it were intended as an outsider's observations, why would he have needed the assistance of Fred Gardner, why would he have needed more from Sam Shepard than translation? (Shepard has said that he put in the scene with the children "and some other scenes.")

What impelled Antonioni to this "impersonation"? It is a commonplace that the most difficult part of an artist's life these days is not to achieve good works or recognition but to have a career: to live a life in art, all through one's life. Since the beginning of the Romantic era and the rise of subjectivism, the use of synthesis in art—of selecting from both observation and direct experience, then imaginatively rearranging the results—has declined among serious artists, until by now art has taken on some aspects of talented diary-keeping. (The most obvious examples are "confessional" poetry and "action" painting.) An artist's life and internal experience have become more and more circumscribedly his subject matter, and his willingness to stay within them has become almost a touchstone of his validity. This has led to the familiar phenomenon of the quick depletion of resources—all those interesting first and second works, and the sad, straggling works that follow them. The question is further complicated because the more sensitive a man is, the more affected he is in our time by the Great Boyg, the presence of the monster, which increases his sense of helplessness, of inability to deal with such experience as he does have.

But he must work, because for him working and living are congruent. He is (in this case) in his mid-fifties, in supreme technical command of his medium. At the moment, let us say, he feels drained of inner resources. This feeling of attrition is emphasized by contrast with some people around him who are not drained and who, artists or not, have vitality and address. They are the radical, dissident young of the world. They make him envious. Not so much to be young again, or to be as virile as he was at twenty-one—those envies apply to anyone. He is mainly envious of their surety and moral energy. He is experienced enough in politics to doubt salvation through politics; he is acute enough morally to see the dubiousness of youthful moral absolutism; nevertheless, he is envious of their *beings*. And, possibly, he tells himself that what he needs is to shift from a generally pessimistic view—which has been the fundamental view of most serious artists in the last century and a half—to a participatory and expectant view, at least as a motor device. His first move is to *Blow-Up,* in which the protagonist and the ambience are young but the center of which is mature, a view down the perspective of some years. With *Zabriskie Point,* the exponents *and* the center are intended to be young.

Helping to expedite this shift is the special relation between this artist's medium—film—and the young. There is no art that excites them as much, no art that seems to belong more surely to them and their era. A lot of young people's attitudes and assumptions are culturally questionable—and potentially harmful to film itself—but, incontrovertibly, those attitudes exist. So the temptation for a film director, of all artists, to take youth not only as subject but as creed is stronger than it might be for a novelist or a dramatist.

Another contributing factor follows from this: Antonioni's success and the pressures that attend film success. The problem here is not in declining to do compromised work, although I'm reliably informed that a few years ago Antonioni refused a million dollars to direct someone else's script. The problem is that he is wanted, that the production money which most directors find difficult to get is his almost for the asking. This is a strong drug for any director, particularly for a man who spent many years starving and scratching and hacking and begging, in order to get the chance to work as he pleased. At last he has power, and he is pressed to use that power by people who now want "product" from him. Here are some possible consequences: he works before he is ready to, and he spends more money

than he needs to. And perhaps by an unconscious chain of association, he proceeds to the place from which the money stems, as the
source of both empowerment and conflict; and, to display fidelity
to his principles, he makes a picture attacking the system that produced the money with which he is making the attack.

Political action has not been a real concern of Antonioni's films
since *Il Grido*. (The strikers in *Red Desert*, the antibomb marchers
in *Blow-Up*, were only elements of the environment.) For a man
of his philosophical temperament—I say this carefully—politics
seemed to have become increasingly superficial, at least as a subject
in art. Now his use of the rhetoric of revolution seems an escape, an
emollient for a fever of frustration, a way of *seeming* to come face
to face with root troubles in men, something that only the best political philosophers can do through politics and for which political
activists can't afford the subtlety, even when they have it at their disposal.

With at least some glimmer of the need for political change in our
society, I still suggest that, for Antonioni, the political gesture of
this film may not be much more than a kind of personal therapy. It
may be an attempt to exorcise a guilt for having been politically
quiescent in his recent work, just as the whole film may be an attempt to exorcise a guilt about being middle-aged. Assuming that
these guilts exist, I suggest that the first one is irrelevant in an artist
of his depth, and the second is a kind of senility in reverse, a contradiction of his own genuine and appropriate vigor.

All the above, I think, are matters that may have contributed to
the mistake of Antonioni's new film. What will happen now? I hope
he returns to Italy, to work. Besides the obvious fact of the familiar
language, there is the para-language, the things that need not be
said between people who understand each other easily; and he will
also understand *societal* language again. I need not repeat how
beautifully I think he has used that language in his Italian films. And
again he will have the chance to be alone, to reside within himself,
something that is very difficult for a man in a foreign country, who
usually feels isolated and ignorant unless he is continually trying to
join.

Youth may continue to interest him. Why shouldn't it? But if he
sees it through his own eyes, out of the core of his own society, he will
have more to say to all of us, including Americans, than by trying

to be a surrogate American youth. How badly we need the best of Antonioni. On past evidence, all he needs to do is to be Antonioni again, unafraid of his doubts and his solitude. In the career of this superlative artist, *Zabriskie Point* can only be the occasion for gravest concern and affectionate hope.

POSTSCRIPT. This review drew a number of disagreeing letters from young readers, who argued for the film in terms of its subject matter. These letters reminded me that there really are such young people as Mark and Daria, that alienation and dissent are all that is possible for them, that nonactors symbolize the nonparticipation of such young people, that there are similarities between young people in Los Angeles and Rome. These letters depressed me, for two reasons. First, in replying to them, I had once more to go over what I take to be the failure of an artist of major stature. Second, they heightened my concern that some young people are so anxious to find evidence of dissent in films that they magnify the virtues of any film that deals with the subject. This latter leads to the, I think, still enormously important matter of the corruption of taste, which means the corruption of values and thus the debasement of the very humanity that concerns these young people. Further, if any picture favoring young radicals or dropouts is welcome, this opens doors wide for the most opportunistic and cynical of Establishment exploiters. Most important, it leads to myopia about the validity of dissent itself.

Luchino Visconti committed an intellectual and political disservice in *The Damned* by implying that all fascists are either money-mad maniacs or drooling sexual perverts; this is not the truth, and thus it is not the danger of fascism, nor its warning sign. If, in films like *Zabriskie Point,* we take superficial business cartoons as the truth about materialism, if we take sloganeering and attitudinizing as the truth about dissent, we blind ourselves to the real sterilities of the first and the real values of the second.

Winter Wind

(*March 28, 1970*)

Winter Wind is confusing, and I left the theater disappointed in this new film by a director I admire, Miklós Jancsó. But then it began to work, retroactively. The confusions are not clarified now, but a rationale for the confusions becomes clear, an esthetic rationale.

This is Jancsó's third film since *The Red and the White* (1968); the two intervening films—*Silence and Cry* and *Ah! Ça Ira*—have not yet been shown here, so it is impossible to discuss his development in an orderly way. But some connections are evident even on the other side of this jump. His theme is once again power, the power of some men over other men's lives, and the desire of the other men to reverse that power structure, and the way that the contest gets transmuted into statements of ideals. Once again he is concerned with the record of those contests, called political history. Once again he uses sex—not love, but sex, nakedly represented by female nudity—as a chemical in the compound. Once again violent death is treated, both by killer and killed, as an event whose time has come; once again fear is never shown, not because all these people are brave but because the characters have accepted their destinies in life, as the actors have accepted playing those characters.

The opening consists of clips from a famous newsreel of political assassination, the shooting of King Alexander of Yugoslavia at Marseilles in 1934 (and also of Barthou, the French foreign minister, who was welcoming him to France). This is shown silently, accompanied only by slow gong strokes. Then we are told that the story we are about to see deals with events leading up to this assassination, which was planned by the Ustashi. The latter were a group of Croatian fascist terrorists, trying to wrest Croatia from Serbian domination.

The first sequence that follows is the first of the unexplained incidents, an assassination attempt by a Ustashi band. We never know where the attempt takes place or who is in the carriage coming along the snowy road. All we know is that the sequence is filmed beautifully, almost idly—all we hear are hoofs and shots, and the cawing of crows, as the camera circles continuously; following the men as they ride and walk across the snow-covered fields, take their places, then fire and are fired upon, and fall or flee, in a waking dream of killing. When Jacques Charrier, as the leader of the band, is waiting behind a tree for the carriage and crosses himself with a hand that holds a pistol, he epitomizes the film's atmosphere.

All the rest takes place on an estate in southern Hungary near the Croatian border, where the Ustashi are hiding while they plan the assassination of the king. The materials of the picture are the suspicions bred among conspirators by the very act of conspiracy, the corruption of small loyalties within the outlines of a large loyalty, the masquerades of ego and sadism under the name of fervor for freedom. The action centers on the leader, Charrier. At the end his followers form an agreement of cooperation with their Hungarian hosts. He refuses to abide by this agreement, refuses to commit suicide, expects to be shot, and is promptly shot. Then he is made into a martyr; the last scene is an oath of fealty to the movement in his name.

That much is clear. But dozens of details in the film are unexplained, abruptly introduced and dropped. Some examples: Why, when the spy is discovered, is he taken out and made to lie down in the snow, then ordered to get up again, and only much later shot? Who is the tall moustached man who brings the teen-age girl with him? Why does he make the two other women in the house undress? Why does he make the girl take a bath with one of them? I have no idea. (There are also anomalies of language: for example, they all speak French—this is a Franco-Hungarian coproduction—yet sing their anthem in Croatian.) Trying to follow the story of this picture, as story, is an exercise in frustration. Yet, incredibly, after the film is over, a feeling begins to seep through the accumulated frustration: that the bafflement was expected by Jancsó; that it is the *choice* of an obviously skilled and accomplished artist.

Subsequently I read a recent interview with Jancsó, conducted by Jean-Louis Comolli and Michel Delahaye, in which they asked him why he refuses to give explanations, why he shows only unmotivated acts. He replied that it is because of conditions in Hungary, where "we cannot put our message across as clearly and directly as we would like. . . . Since all of us want to say something, and at the same time are not forthright enough to do so, we count a great deal on the form of the film to make a statement. . . . Perhaps that is the reason why situations and acts are the most important things in my films. What interests me most is form."

He says that for several centuries, because of continually changing internal conditions, Hungarians have had to speak obliquely, using shape to suggest substance. We might cite exceptions, but perhaps Jancsó has found a rationale in the present authoritarian Hungarian state for a "private" method that he really has arrived at out of his own psychology and esthetics. At any rate, his film does find its life through its form.

This form has two principal aspects. First, the visible statement of unexplained acts in order to reach the unstated. The aim seems to be to make a film with the *aftereffect* principally in mind. Second, the use of continuous, generally circular camera movement. There are fewer than fifteen cuts in this film: which means that the entire eighty minutes were photographed on fewer than fifteen pieces of film. Compare that with, for instance, *Rashomon,* which runs eighty-eight minutes and is composed of 407 bits of film. Jancsó literally *follows* the action. Kurosawa has said that he photographs in order to have something to edit. Jancsó is photographing in such a way that editing is superfluous: which means that a good deal of his creative process takes place at an earlier stage in the making of his film.

Also, he has said that he shoots his films silent, or with a sound track merely for purposes of record, so that he can direct his actors while the camera is rolling and moving. (The final sound track is added later, usually with the actors who have played the roles.) This gives him the chance to keep some sequences running more than ten minutes and, although he has rehearsed them before he shoots, enables him to "conduct" the actors, à la silent film days—or just as a conductor leads musicians whom he has already rehearsed. These long sequences, without cuts and with the flow of the camera's movement, are, I think, Jancsó's effort to capture some of the long, relentless, and cyclical swell of history, thus reinforcing his theme with the look and feel of his film.

Theory is theory: and some excellent theories, excellently articulated—like Brecht's theory of epic theater—simply do not function as well as they read. But I am haunted by a feeling that Jancsó is on to something fascinating, imperfectly though he has accomplished it here: an attempt to record the subliminal—not with the much-used method of splattering in bits of nonsequential or non-"present" material, but by trying to capture a mystery beneath the surfaces of facts. A chief power of film has been its ability to make facts magical;

Jancsó is trying to ignore that magic, is trying to catch the invisible *sum* of the facts, which is like the solid wheel that is made by spinning individual spokes. His inquiry is both thoroughly cinematic and quite revolutionary, but in a sophisticated and difficult way—much more daring than the Now editing that has been diluted by Now hacks out of Godard and Lester.

But, as noted, *Winter Wind* is unsatisfactory—possibly because it does not dare quite enough. A ballet is a ballet: we do not expect character dossiers or precise explications of this glide or that leap. But Jancsó has paused at a midpoint between veristic drama and non-programmatic ballet. A film on the *idea* of politics that, out of all history, specifies the Ustashi and then does not tell us why, has one foot in realism and the other in abstraction. Was Jancsó trying to say something about Hungary's dictatorial government of the time and its support of fascists? Was he trying to show us ironically that fascists are as pure in their revolutionary zeal as other zealots? These are only some of the foreground questions that he himself raises, then ignores as he concentrates on abstract matters of form and movement.

Charrier, in past films, has been an unremarkable good-looking young man. Here he manages more gravity than I would have expected but not quite enough broodiness and concealed fire. Marina Vlady suggests some sexual secrecies, as a woman assigned to the hideout. Zoltan Farkas, who has edited much of Jancsó's past work, is again his editor here, but, outside of possible help on planning the shape of sequences, could not have had much to do. This is the first color film by Jancsó I have seen, with the cinematography by Janos Kende, and the palette is generally muted.

The matters I noted in *The Red and the White*—the constant shifts in authority from one person to another, the strings of footling commands—have now grown almost into raisons d'être, along with longer and longer camera sweeps, and the result is an irritating, unsatisfying picture; but *Winter Wind* is the work of a unique and complex cinematic mind.

Fellini Satyricon

(April 11, 1970)

FEDERICO Fellini told Alberto Moravia, in an interview lately published in *Vogue,* that he had tried to eliminate the idea of history from his *Satyricon,* "the idea that the ancient world *really* existed. . . . I used an iconography that has the allusiveness and intangibility of dreams." In reply to the next, logical question, he said that it was a dream dreamed by himself, and then Moravia asked, "I wonder why you dreamt such a dream." Fellini replied: "The movies wanted me to."

Exactly. The reply contains all the truth and fakery and truth about fakery that have made Fellini, artist and man, one of the most appealing of modern film figures. Since his earliest films, *Variety Lights* and *The White Sheik,* he has been dealing with truth tellers and pretenders, realists and dreamers, and his love for both. *I Vitelloni* is about some young men who are realists and others who are fantasists. *8½* is about a film director begging the movies to command a dream from him. Fellini's life has been spent in the service of both reality and nonreality, largely because he knows, as one of the few film masters who also understand theatricality, that theater without artifice is a fake ideal and a naïf's idea of truth.

His movie dream of Petronius is another work of truth and artifice. I had to see the picture twice in order to get some view of it; the first time I laid my own expectations on it and obscured the work. I expected (and the advance publicity whetted the expectation) a comic bacchanalian film in the tone of the original, which has been described by William Arrowsmith as "everywhere shot through with a gusto and a verve and a grace of humor." Not at all. *Fellini Satyricon* (so called to distinguish it from a quickie competitor based on the same source) is elegiac, joyless, resigned. There are many scenes of revel and of sex in it; there is very little gusto.

Another burden from which Fellini has to be freed is our expectation of method. He has taught us to expect lightning play in his editing, swift referential humor and counterpoint, drama and dialectic by deft junctures of material, and he has used this method even in his recent short film *Toby Dammit* (a part of *Spirits of the Dead*). There is some splintery referential editing in *Satyricon,* but the principal method is immersion in texture and color, steady progression through the "feel" of a scene, rather than any lightning mosaics or kaleidoscopic flow. The second time it was possible to see the picture as what it is.

Fellini and Bernardino Zapponi, who also collaborated on the screenplay of *Toby Dammit,* have used enough of Petronius to bind their script to the original, in materials at least, and have also done some improvisation on themes. We see the rivalry between Encolpius and Ascyltus for the pretty boy Giton. We go to Trimalchio's feast. We visit bordellos and baths. We follow the two friends on their adventures as slaves and freebooters. There is a quiet episode in which a patrician commits suicide to avoid capture by an enemy, and as he sits tranquilly bleeding, his wife, at his feet, murmurs (anachronistically) the opening of the Emperor Hadrian's lovely poem *Animula vagula blandula.* Encolpius and Ascyltus enjoy the favors of the patrician's sole survivor, an African slave girl whose language is like running water. Encolpius later loses his sexual potency and journeys with Ascyltus to a witch who can restore his virility. While Encolpius is delighting in his returned maleness, his friend gets involved in a fight for his life, which we see in the background. But Encolpius does not leave off to help his friend, and Ascyltus is killed. Then Encolpius embarks on a voyage with new friends. He begins to describe the voyage for us but breaks off in midsentence. The camera pulls back from the seaside to show us fragments of frescoes bearing images of the characters we have been watching for two hours, and the film is finished.

Fellini has often used multinational casts but always with some eye to seeming Italian-ness. No such worry here. His actors are American and English and French and German, as well as Italian, and the sound track is a mélange of languages—Latin and Greek and Italian and German—which he has dubbed with almost deliberate carelessness, as if the Voices, collectively, were an additional character. He says he has done this for an effect of alienation, and he has succeeded, for good or ill. Part of the result is that none of the people achieves anything more than time-experience with us; we get no deeper knowledge of them, only more and more of the first impression.

Some of the faces are excellent: Martin Potter, the golden Encolpius, Hiram Keller, the dark scheming Ascyltus, Max Born as Giton, the faun of many an afternoon and night, Mario Romagnoli, a Roman restaurateur with the perfect parvenu face for Trimalchio. Among the few familiar actors are Alain Cuny, as the one-eyed Lichas, and, in brief appearances, Capucine, Magali Noel, and Lucia Bosè. There is the usual gallery of Fellini "faces," mostly grotesque; but they are not used as English social-realist directors use ugly faces, to prove that life is ugly: they help to fill out Fellini's theorem that the gods will have their little jokes and that *all* faces, including pretty ones, are among the keenest jokes, one way or another, eventually.

The cinematographer, whose lighting makes the evocative murk or glare, is the masterly Giuseppe Rotunno, but the visual base of the film is its sets and costumes by Danilo Donati (in place of Fellini's usual designer, Piero Gherardi). The intent of the costumes here is quite different from that in *8½* and *Juliet of the Spirits*. There we saw extravaganzas on clothes that we know; here that view is not nearly so possible, and what we get is warped sumptuousness—a lushness that is not meant to attract.

This leads directly to the matter of Fellini's purpose in *Satyricon,* so far as it can be inferred. I can't place much credence in the formulations that he himself has offered in incessant interviews: that this is a pre-Christian film for a post-Christian era, or that it is science fiction of the past instead of the future. This cautionary reason may have some meager basis *ex post facto,* but it hardly seems large enough. To show us parallels in decadence? For all his love of superficiality, Fellini is not as superficial as that. True, during Trimalchio's feast, I remembered that one of Scott Fitzgerald's projected titles for *The Great Gatsby* was *Trimalchio in West Egg;* still the modern parallels are either obvious or strained. For instance, homosexuality in Petronius is part of the norm, but the most unbiased among us today knows that homosexuality is generally considered perverted or sinful; so Encolpius' love for Giton does not symbolize modern "dissolution." Fellini's reasons, I think, lie elsewhere. Moravia has recently reviewed the film (*New York Review of Books,* March 26) and after a couple of columns of banality on Petronius—compare Erich Auerbach!—he gets to what he considers the content of the film, which is "broadly speaking, religious. . . . To understand Fellini's special kind of religiosity, we believe that the greatest importance must be attributed to his manifest preference for the mon-

strous and impure. . . . One does not have to make a great effort of
the imagination to trace back this preference . . . to a fascinated and
funereal moralism." He then says that the difference between Petronius and Fellini is that the former is a realist and the latter is not.
Auerbach (in *Mimesis*) agrees about Petronius only within strict
historical limitations; I would agree about Fellini only with regard to
his *Satyricon;* but in any event Moravia's analysis leaves unclear any
internal reason, inferable by the viewer, as to why Fellini made this
film. In short, what is there—in the picture itself—that indicates
why this man, who has made only contemporary films that were
psychologically pertinent even when stylistically extravagant, has
abandoned pertinence for extravagance: has chosen a subject that
freed him of pertinence and allowed him to concentrate on the
extravagance?

For me, the answer lies in flat contradiction of Moravia's statement that Fellini has "completely surmounted the personal crisis"
with which his two previous full-length films were concerned. *Satyricon* wrestles with that crisis in another way, that is all. And that crisis
makes a connection, at the very base, between this *Satyricon* and
Antonioni's *Zabriskie Point.* Both are the works of mature artists that
reflect the contemporary artist's relation to the world as material for
art. Experience is not less than it was, it is too much more: our
culture's expanded consciousness (within) and amplified communication (without) overwhelm and enervate some artists and produce,
finally, a bankruptcy, rather than a surfeit, because of a sense of in-
competence to deal with that enlarged experience.

Still, artists must work or wither. Antonioni's solution, as discussed earlier, was an emigration to a different place and a different
generation. Fellini's emigration was to the past: where his sense of
futility and oppression was relieved of the necessity of point and *could
express itself as a function of film making itself.* Moravia is quite
right about Fellini's moralism, but one large aspect of that moralism
is in the artist's moral compunction to work. It is difficult—at least
for me—to imagine Fellini making this film unless, in a way, he was
forced to. *Satyricon* is a step past *8½,* which was about a director
looking for a film to make and (despite the desperate ending) failing
to find one. *Satyricon* is the film that Guido, the hero of *8½,* might
have made.

The statement that Moravia himself recorded in his interview seems to me the root of the matter: the movies commanded Fellini to dream this dream because he is alive and his life consists of making films. It does indeed deal with the monstrous and impure; its moral tone is funereal. It might better have been called *Fellini Inferno,* rather than his *Satyricon.* But the inferno, I believe, is the sum of the conditions of life, and *his* life in particular, that forced him to make the film at all.

So the film depends for its being entirely on the way it is made. There are of course recognizable Fellini hallmarks: the silent opening (as in *8½*), the big fish (*La Dolce Vita*), the abrupt ending (like the freeze frame at the end of *I Vitelloni*), the earth mother whore (from several pictures). But it is the first Fellini feature film that has, in the post-Renaissance sense, no characters. There are only persons, some of whom are on screen more than others. The film has no cumulative story, let alone drama. There is not even a cumulation of adventures, in the picaresque manner; many of the sequences are simply scenes observed. *Satyricon* depends entirely on its look, and, unlike *8½,* which finally lives through its style, there are few afferents to bind us to the style, to make us care about it in anything more than a graphic arts, "gallery" way—a way that is directly opposed to theatrical experience.

Fundamentally we simply look at Fellini's pictures, some of which are gorgeous, some rather predictably so, all of them tending toward darkness and away from light. (Sunlight actually kills one of the persons in the film and tortures some others.) Sex consists of lurid flickering lights in that darkness, whether they are glimpses of bright cubicles in a bordello or torches that magically ignite between a beautiful girl's legs. The scaly, scumbled walls, the faces that are fighting mortality, the pleasureless pleasures, the short-lived parodies of friendship and love and marriage, all of them are essentially pictorial, not dramatic—pictures from the dream of a man who once had exuberance and now has only dreams. It is the work of a master who knows, rather somberly, that he still has a long time to live.

Mississippi Mermaid

(May 2, 1970)

Any film by François Truffaut commands a certain amount of ecstasy in advance from those committed to the *auteur* theory of film criticism. One feels that their response to all future Truffaut films is already pretty well formulated and awaits only details of plot.

For those uncommitted to that theory, Truffaut's *Mississippi Mermaid* may seem—as it did to me—a commonplace, overextended thriller. As is often the case with Truffaut and others of the old New Wave, it is based on a hack American novel—this one by Cornell Woolrich. Jean-Paul Belmondo is a rich young factory owner on the island of Réunion, near Madagascar, who sends for a mail-order bride. She is Catherine Deneuve, who arrives on the ship *Mississippi,* hence the title. Within about thirty seconds, all of us except Belmondo know that Miss Deneuve is an impostor, not the girl with whom he had corresponded. About forty minutes of screen time later, Belmondo, too, discovers the deceit, and the rest of the film is taken up with his pursuit of the decamped deceiver, who has robbed him, his acceptance of her (when he finds her) because he cannot help loving her, his commission of a murder, and the sad, long-delayed end.

In all of this decently photographed and passably acted film, there is no plot surprise, except for the very clumsy discovery of Belmondo's buried victim. Nor is there a great deal of the visual felicity that was once Truffaut's distinction. Belmondo walks through his performance like a latter-day Jean Gabin, although, like Gabin, his mere presence has some power. Catherine Deneuve, blank-faced and skinny, is again supposed to suggest, by the negatives of her blankness and skinniness, all sorts of fire and kinky sex. Michel Bouquet, tiny-eyed and purse-mouthed, is adequate as a private detective.

As with so many films by directors of this school, there are lots of "in" film references—called, in the lingo, "homage." The picture is dedicated to Jean Renoir, but that at least is overt. Some cryptic references: the detective is named Comolli (Jean-Louis Comolli is a well-known French film critic); the house in "Switzerland" to which the couple flee at the end seems to be the same house used at the end of Truffaut's earlier (and immensely superior) film *Shoot the Piano*

Player; there are matte shots à la silent films (where all of the screen is masked except for one small area) and "wipes" between scenes. These last devices are meant to be seen in invisible "quotation marks," as when a literate person deliberately uses clichés or archaicisms.

For those uninterested in "in" cuteness, these deliberate clichés are tedious, and the unwitting clichés make them worse. For instance, the sequence in which Belmondo is sick in a Nice hospital might have been shot by such a Hollywood dray horse as Jack Smight.

Truffaut was a leading formulator of the *auteur* theory; in fact, as explicit theory, it is usually said to date from an article he wrote in *Cahiers du Cinéma* in January 1954. One tenet of that theory holds that material is less important than its cinema treatment, thus these directors have often taken stock genre material, like American thrillers, in order to prove that film art can be made out of the film's "own." Sometimes (as in *Shoot the Piano Player*), the transmutation succeeds; more often, the result is only a combination of smugness and camp, accompanied in the theater by the purring of the viewer who gets the "in" references and relishes the exaltation of pop over pompous old "elitist" art.

When the transmutation fails, as it does in *Mississippi Mermaid* and as it does in most cases, the *auteur* theory shivers. André Bazin, the late French critic who was early associated with the theory and who was Truffaut's mentor, wrote a corrective article in 1957 which is generally disregarded by his disciples. Bazin said: "All that [the *auteur* supporters] want to retain in the equation *auteur plus subject* = *work* is the *auteur,* while the subject is reduced to zero. . . . *Auteur,* yes, but what *of?*" The answer, in regard to Truffaut's recent films, is: Not much.

Brotherly Love

(May 9, 1970)

A hopeless note. People won't go to see a fine performance in a poor film, as they did not go to see Richard Burton in *Staircase,* but it would be derelict not to notice fine work. *Brotherly Love* is a superfluous film. The director, J. Lee Thompson, is a pedestrian cinematic mind (though he evidently understands something about acting). The late James Kennaway, who wrote the script from his own play and novel, was a very diluted Brontë. But Peter O'Toole's performance is stunning.

The title is literal—the story is about incestuous love (not acts of incest). The theme is ancient, but, to supply a double and symbolic doom, it is intertwined here with a theme of declining aristocracy, which makes it somewhat remote for us. O'Toole is therefore working uphill all the way, with little base of reference in American life for his character's nostalgia, yet he succeeds brilliantly in creating an affecting lost man.

He is a Scottish country baronet, sensitive, arrogant, alcoholic, who loves his sister (Susannah York) so much that her husband has left the house, though he lives nearby and wants his wife back. O'Toole is in process of ruining himself physically and financially as a rather theatrical Last of the Line, a detester of plebs, and a lover who knows that the one true love of his life is hopeless. His sister returns his love almost as completely, and when he sees that he is ruining her life, driving her to sordid affairs in her helplessness, he goes pretty much out of his mind—a place he has been before.

O'Toole has to survive some coarse incidents, like a tumble among garbage cans and chamber pots to prove how low he has fallen. But from his first appearance, there is conviction that a very difficult note has been struck *exactly,* with subtlety and fire. Through all of this easily capsizable role, he balances pride and self-disgust, amusement and anguish. His voice is one of the best instruments among English-speaking actors, not big and boomy but delicately colored and strong. When the drunken brother tells the husband that the latter really hates them (the brother and sister) but is hooked on them, O'Toole charges the lines with the electricity and music of damnation. When he begins to go feral and shifty and suspects even his sister of conspiracy against him, his gaunt figure has a lone Byronic appeal. At the end he flees those who want to cart him off to an asylum, and his tearful sister finds him in their old hideaway in a barn; and when he turns to her in the dim light and speaks the words "I love you" for the first and probably last time, it is a moment of high pathetic beauty.

But it's all no use, I know. Not many people are concerned about acting, as such, any more. Worse, O'Toole's curse is that he is a romantic actor, in the largest art sense of the term, caught in an age that has little use for that kind of romance. "It's a sort of time thing," he says at one point, "I'm what's known as a hangover." It is a performance of an anachronistic character in what is perhaps becoming an anachronistic style, but it is splendid. *Will* anyone see it?

Two or Three Things I Know about Her

(May 9, 1970)

IT is now chic to speak of Jean-Luc Godard's Periods. The Second Period presumably begins with *La Chinoise* (1967) and concentrates on political-revolutionary themes. The First Period, though heavily laced with politics, concentrated on personal relationships, societal matters, and investigations of artistic method. Now, belatedly, we get the last picture of the first group, *Two or Three Things I Know about Her,* made in 1966 and seen here only at festivals. It turns out to be one of Godard's best.

The "her" of the title, he says, is not the heroine but Paris itself. Two events impelled the making of the film: he read a newspaper letter from a woman about part-time prostitution in the new expensive apartment projects because of the high cost of living, and in August 1966 a new administrator in chief was appointed for the Paris region. We see a day in the life of a young wife, including afternoon rendezvous to pay for a new sweater. Her story is frequently intercut with the sights and strident sounds of changing Paris: drills, cranes, construction trucks, etc. Near the end she stands in front of the apartment house where she lives with her garage-mechanic husband and their two small children, and as she talks to us of her plight, the camera does a Godardian 360-degree pan around the immense canyon of new houses. The very last shot is of neatly arranged package goods on the grass—detergents and cigarettes and breakfast foods—and they glow at us as the picture fades.

The aim is not solely to mock materialism: Godard is also influenced, reportedly, by the poetry of Francis Ponge, which is concerned with "Thingism," the seeming life and effect of "things." But the film mainly explores the Americanization (read "modernization") of French life and the pressures of artificially stimulated consumerism, which, Godard implies, makes prostitutes of us all to some extent. His cinematic techniques are familiar: no plot but a chronicle with Brechtian part titles; many interviews; some literary references. (Two young men, who are seen compiling an anthology of unrelated excerpts from books, are named for Flaubert's Bouvard and Pécuchet.) A commentary is whispered throughout by Godard himself, including quotations (he has said) from Raymond Aron's *Eighteen Lessons on Industrial Society.* There are many pretty girls and much cigarette smoking as décor. The color photography by Raoul Coutard is, as usual, exemplary.

Two or Three Things is more interesting than many other Godard films because, for one reason, it seems to have sustained the director's own interest. There is no feeling, as in *Pierrot le Fou,* that this very bright man has embarked on something to which he is committed long after his darting mind has really left it and that he has been forced to invent irreverences and interpolations to keep himself interested. For another reason, the film is devoid of the worst aspects of Youth Worship that sometimes taint his work; it is about people, some of whom are young. But the chief merit is that it develops its themes within itself, for the most part, not by imposition. The interplay between the facts of the changing city (an appeal to the new Paris administrator, really) and the changing lives of Paris is graphic. And when the heroine moves easily from action within a scene to speak to us directly about herself and her quandaries, which she does often, it creates two dualities of consciousness—hers about her life and her "acting" of it, ours about the film as fictional truth and about the making of that truth. There is a nice sense of metatheater, in Lionel Abel's term: of the heroine living her life and simultaneously seeing herself, as the protagonist of a drama she is watching. And all the while, a tightening circle of chromium-plated, electronic wolves is yapping at her heels.

But the impasted artistic and philosophical freight is once again tedious. The interviewing of characters by an unseen interviewer, which is supposed to break open film convention, is now a Godardian convention. The sound track, with Godard quoting away, has an air of dormitory discovery—a sophomore discovering, under the midnight lamp, what life and metaphor are All About. When we get a huge close-up of bubbles floating on the surface of coffee in a cup while Godard whispers about Being and Nothingness, it remains bubbles and quotations; there is no transformation into philosophical comment or Pongeist poem.

Marina Vlady, the heroine, is pleasant and composed. While the film stays with her, in her complications of self-knowledge, there is some sense of genuine phenomenological dilemma, some inquiry into the data of consciousness. When Godard sloshes *stuff* at us, belatedly discovered by him and untransmuted, we get a Child's Garden of Phenomenology.

The Joke

(May 16, 1970)

HISTORY has supplied a nasty final twist for *The Joke*. This Czech film has its own bitter finish, but it is now topped by events in the world around it. In the mid-1960s Czechoslovakia was looking back with bitter relief at the Stalinist days of the early fifties. The hero of this film is a science researcher who suffered in his student days because of a joking postcard he wrote, mocking official optimism and praising Trotsky. A student committee got him expelled from the university and into a punishment brigade. Now he seeks revenge by seducing the wife of the chairman of that student committee. After the seduction, the joke is still on the hero because he finds that the couple are estranged and he hasn't hurt the husband at all.

End of film. But what has happened in Czechoslovakia since it was finished takes the point past irony into the mainstream of historical pessimism.

Jaromil Jires, the director, collaborated with Milan Kundera on the screenplay, derived from Kundera's novel of the same name. Jires was not making a satire on bureaucracy. That sort of satire is a staple of current East European literature, permitted because it is unspecific: it could apply to Madrid or Canberra or Washington. Jires' film is deeper: it is about the difference between a socialism that is fearful and rigid and a socialism that is wanted and therefore is not afraid of mockery. That is Jires' political point. On the human scale, he is saying that revenge is impossible because time cannot be turned back and wrongs cannot really be undone. Revenge may make us feel better, but that's another story.

Josef Somr, who was the Don Juan in *Closely Watched Trains,* is exactly right as the saturnine hero. (One nice touch: when he gets up in the morning and dresses for his vengeful assignation, he doesn't shave!) Ludek Munzar has a clean sanctimonious air as his prosecutor and intended victim. And Vera Kresadlová, the girl in *Intimate Lighting,* is again charming in a brief appearance as Munzar's young mistress. Jires makes a particularly trenchant comment on the generation gap in a brief exchange between Somr and this girl. She is eating cotton candy at a street fair when Munzar introduces her to Somr. The conversation turns to the subject of physical strength, and she teases Somr. He replies that he is very strong, that he used to work in the mines. She takes a bite of candy and asks whether he was in a punishment brigade; he says yes. "How long?" she asks. He says, "Six years." Without a pause she asks, "Where?" and takes another calm bite of cotton candy.

Getting Straight

(May 30, 1970)

Getting Straight is a very ambitious film that is too small for its britches. It wants nothing less than to deal with the educational crisis, the student crisis (not precisely the same), the racial crisis, the generation gap, the draft, Vietnam, drugs, relations between the sexes, and sexual relations. The screenplay by Robert Kaufman, from a novel by Ken Kolb, scurries after all these matters on one of the busiest shopping tours in recent films. Of course, no film about university life today could easily avoid at least some of these subjects, and *Getting Straight* touches at least some speck of truth in every one of them, but its movie-type cleverness and its perspiration as it hurries after everything reduce the impact of the best in it.

If this picture hasn't already been nicknamed *The Postgraduate,* it will be, and the joke is justified—although it has nothing like the virtues of *The Graduate*—because it proves yet again that Nichols' film was a milestone in U.S. film history: *Getting Straight* traces its pedigree all the way back through those three long years to Nichols. The postgraduate himself is Elliott Gould, returned to a Western university six years after he got his B.A.—having spent some of the interim in Vietnam and some just copping out—to get his M.A. and teach in high school. He means to get "straight," to join up and join in, but he ends up, after some fury at the blindness of the Establishment, by joining the student dissenters.

The first few minutes tell us, in stock vignette style, that Gould is a Big Man on campus, everybody's friend, on top of every situation and a great many girls. He is humane, broke, sincere, intelligent, hard-working, fiery, sexually gymnastic, undeluded, idealistic, and witty. Burne-Jones would not have dared to paint Galahad in pre-Raphaelite light the way this film paints the mod Gould. He commits only one mistake in the story—a fraud, allowing a friend to take an exam for him and forge his name—and even that is so egregiously dumb that we know it was planted by the scriptwriter just to provide one personal flaw and a plot complication.

There are intellectual sentimentalities: Gould, as a practice teacher, tells a slow student who reads *Batman* that this is great because it leads to *Don Quixote*. There are structural pangs: the quarrels between Gould and his "real" girl, Candice Bergen, pump away in the script like an artificial heart. (The two finally come together in the middle of a wildly destructive student riot—like any old fade-out clinch that made a quiet oasis in the middle of a storm or a battle.) There are dreadful gags: some mechanical stuff with Gould's battered car. There is factual silliness: the liberal university president is gleeful about "revolutionary" concessions he has wrung from his trustees, concessions that would have seemed outdated in 1965. (Not that there aren't reactionary campuses still, but those that are liberal are far more liberal than this president's program.)

Still the film has some bite—particularly in a few of Gould's outbursts about the blindness of the Establishment and the need for revised values. Enough bite so that this picture raises questions bigger than its treatment of them. *Guess Who's Coming to Dinner* (for instance) was too stupid to make us think about its subject: *Getting Straight* is just intelligent enough to make us think of what it's about: the function of the university, its relation to the festering world that supports it, its relation to the students whose response to the university's education is to quarrel with the university—basically, questions that have been troubling our world in the hundred years beginning with Matthew Arnold's *Culture and Anarchy*. Culture is still connected to the evils that induce the rage to anarchy, and now the potential anarchists come from among the cultivated. It is an old joke about Americans that we think every question must have an answer, but we seem to be learning otherwise in these matters. More and more it seems that the best way to deal with these questions is simply to bear them constantly in mind, to deal with them on the best possible day-to-day empirical basis, and to hope that the shape of one's psyche and intent will eventually shape a pattern—though not an Answer.

Getting Straight at least dramatizes—or cinematizes—some of the relevant evidence. Richard Rush, who has made films previously but whose work is new to me, has directed at about the level of Peerce's *Goodbye, Columbus:* lots of physical closeness, juicy color, fast cutting, and multiplane composition. With this latter, he vastly overuses the device of "racking" camera focus in the middle of a shot—a foreground figure becoming misty as a background figure becomes clear, or vice versa; and his sense of bustle seems to invade even the quiet scenes. His film is much more exploitative than committed, but some of the exploitation—like a student battle with cops—is well handled. Laszlo Kovacs, who has worked with Rush before and who also did *Easy Rider,* supplies good post-*Red Desert* photography.

Elliott Gould's forte is comic composure and, within it, the Feifferlike cartooning he can do with inflections and pauses. He gets little chance for it here; he seems to be shouting and "energizing" a good deal of the time. He's solid and pleasant enough, but better in roles like the doctor in *M*A*S*H*. Candice Bergen, presumably whipped along by Rush, tries hard for volatility, but there is a band of deadness across her eyes and a residual flatness in her voice that make her seem stolid in the midst of rage. And I have rarely seen so pretty a girl with so little sex appeal. I think it's her basically unconvincing acting that gets in the way. Professional sexpots like Raquel Welch don't pretend, in their roles, to much credibility: their performances are meant as quasi jokes. Miss Bergen's is not, and its dullness affects her appeal as a girl.

Sympathy for the Devil; See You at Mao

(June 6, 1970)

MORE from Godard—two of his more recent films, made in England in English.

Godard is, demonstrably, a puritan, but he understands the role of sex in revolution. *Sympathy for the Devil* begins with a highly sexual figure, Mick Jagger, recording the title song with the Rolling Stones. These recording sessions, full of the rock-Jagger-youth charge, are the base to which the film keeps returning, in rondo form, after interludes in which black men caress and murder white girls, the camera pans slowly over lurid magazines in a porno bookshop that is run in a militaristic manner, a porno political thriller is read on the sound track, and two miniskirted black chicks interview a black militant. If there is a specific theme in this generally revolutionary film, it is the concurrence of sexual and political energy. A girl called Eve Democracy (Anne Wiazemsky) is interviewed in a manner that rambles physically and topically, mostly political-social questions that can be answered yes or no, and even here there are questions about the orgasm and lovemaking.

In other interludes Miss Wiazemsky wanders about London in a digger hat spraying slogans on cars and walls, usually neologisms that are reminiscent of Walter Winchell. ("Cinemarxism" reminded me of "cinemarriage.") There are readings of LeRoi Jones and Eldridge Cleaver and other black revolutionary material in a London automobile graveyard populated by armed and arming blacks. In the last scene, the wounded Miss Wiazemsky—wounded, I suppose by bourgeois society—staggers across a beach in front of spectators and falls across a camera crane. The crane, at Godard's visible order, then lifts her into the sky as the black flag of anarchy and the red flag of revolution flutter on either side of "Democracy's" body. All this material, and more, seems to rise from and tend toward the sexual springs beneath this film. (Originally, the title was *1 + 1*. Against Godard's wishes, the producer put in a complete rendering of the title song, "Sympathy," and used that title for the picture.)

Sympathy should be considered along with Godard's other English film, *See You at Mao,* which runs fifty minutes and which was made for the BBC but was never broadcast, presumably because it contains a long close-up of a girl's pubis. Revolution is the theme of *Mao,* too, though (despite the pubis) with much less emphasis on sex. Its style is similar to that of *Sympathy:* no attempt at narrative, even unconventional Godardian narrative; simply blocks of material, juxtaposed but not joined, with slow camera movements within each block, with some reprises, and with a good deal of disparate "wild" dialogue on the sound track. For instance, the opening sequence of *Mao* is a very slow traveling shot down the assembly line in a sports car factory, something like the traffic-jam sequence in *Weekend.* Through much of the shot, the sound track is unrelated to the factory: it consists of antiphonal readings by a man and a woman of Maoist dicta, interspersed with the voice of a woman teaching a small child some harsh facts of British labor history. This simple disjuncture between eye and ear is evidently supposed to produce complex tension. Other scenes include one of the naked girl strolling about her flat, a Communist workers meeting, a students' meeting, and fists bursting through paper British flags. The sound track frequently reverts to the theme of lies and truth: bourgeois lies and revisionist lies (the Soviet Union, presumably) as against militant (Maoist) truth.

Both films use a collage method that has its antecedents in surrealism, whose essence was to project disconnected material at us in order to surprise and breach the expectations of rationality and thus to reach the truth that lies beyond the "sham" of reason. So there's a paradox. Surrealism would, not long since, have been the last method to be expected from a Marxist revolutionary. As a style, it has had little to do with agitprop or didactic politics or even with the puritanism that so often accompanies revolutionary fervor. Yet for Godard, who is more and more fiercely Maoist, the collage method in both these pictures seems much more temperamentally congenial than narrative, even the free-wheeling narrative of *La Chinoise* and *Weekend.* He seems more comfortable with a method of fragments than in fragmenting the conventional. And the suitability of this new method for Godard adds greatly to the conviction of his political intent in these new films. Political genuineness and this free form seem to have arrived simultaneously in him. Susan Sontag's comment that his didactic statements are "units of sensory and emotional stimulation" seems to be outdated. The ideas, no longer décor, are meant to affect us as ideas, not as mere sensory elements, and are meant to do so through this sophisticated esthetic mode. (The quality of those political ideas is a quite different subject.)

There is nothing in these films about revolution itself that is new. I was intermittently quite bored with both pictures. What was interesting some of the time in immediacy, and all of the time in theory, was that Godard has evolved a propaganda form that is shaped by the so-called New Sensibility—of the rock-television-cinema age. He is expressing pure and venerable Marxist doctrine in a style that is shocking to traditional Marxist esthetics.

The Ballad of Cable Hogue

(June 6, 1970)

SAM Peckinpah's *The Wild Bunch* was a Western that transcended the form: that is, at its best it pierced right to the forces in us that had originally called the form into being. There were elements of this penetration in his previous Westerns, *Ride the High Country* and *Major Dundee* (the cavalry variation). Now, in *The Ballad of Cable Hogue,* Peckinpah has used just the opposite approach: he has concentrated on the form itself. He has tried to make form and style his material, and in my view the result is a disaster.

With his scriptwriters, John Crawford and Edmund Penney, Peckinpah has aimed at creating a ballad, a film that consciously celebrates its lyric tone and idealized statements, but the effort is so self-conscious and cute, so ill-suited for Peckinpah, that the result is a series of smug contrivances. Over an opening song—a bad one, too—an old prospector named Hogue is abandoned by some partners and, almost dead of thirst but refusing stubbornly to die, discovers a waterhole near a stage trail. Through the years he builds that hole into a prosperous stage stop until he is killed—aha!—by one of the first motorcars.

There are two other principal characters: a lecherous fake preacher, one of those carpentered-up eccentrics that recur in pretentious Westerns, played by the English actor David Warner, who, since *Morgan,* has gained neither in charm nor credibility; and a prostitute

with a Heart of Gold, as well as gold-producing parts. She is played
by Stella Stevens, who has very much less appeal than, but all the
talent of, Kim Novak. Jason Robards is Hogue. From the first moment, he is physically phony, because this grizzled desert rat has a set
of teeth that would be too white for a TV denture-cleaner ad. On this
initial phoniness, Robards builds with empty gusto and stale actor's
rhetoric, and turns this intended folk ballad into a ham salad.

Peckinpah helps very little. He was not attempting the reality-past-
realism of *The Wild Bunch,* but this hardly excuses such japery as
having the face on a five-dollar bill wink at us or shoving his camera
down Miss Stevens' bodice or between her legs to hint that Hogue
wants her; nor does it excuse the stereotypes of encounter and surprise, the stilted opening and close, or the general lack of spirit and
flourish that successful artificiality needs.

Anyone can see what this film is supposed to be about: verve versus
viscosity, man versus mass-man, mortality under the timeless sky—
and there is even a hint of the future ecological crisis. But a choice
of good themes is not the same as making a good film, surprising as
that platitude still seems to be to many. Peckinpah is a truly but
narrowly gifted director with an instinct for violence, some conception
of space, and of human character weathering within that immense
space. He handles these matters well when he looks at them grimly,
and I hope he soon returns to the grim vein. In his hands, the light
fantastic is leaden.

The Passion of Anna

(*June 20, 1970*)

INGMAR Bergman's *Shame* had part of its origin, I thought, in Bergman's impulse to use some qualities of his "company" of actors in a
certain situation, and his new film heightens this sense of collaboration
and continuity. He is the only director now working whose new films
grow organically out of his preceding ones: in continuity of locale, of
associates, of almost diaristic closeness to the director's inner experience. It becomes increasingly plain that Bergman wants this continuity
to be seen as part of his esthetics.

He indicates this in several ways besides the ones mentioned above. There are recurrent names among his characters. Just one example from his latest film, *The Passion of Anna:* Vergérus, the architect, echoes Dr. Vergérus in *The Magician.* There are "quotations" from past work: a dream in this new film looks like an out-take from *Shame.* And there are subtler references: the camera fixes for several minutes on Liv Ullmann telling a story, as it did in *Hour of the Wolf* and *Shame;* the final fade is to white, and *Persona,* after the titles, began by fading in from white.

What is all this continuity in aid of? Since 1961, ever since he settled on Sven Nykvist as his cinematographer, Bergman has made eight feature films. With the exception of the disastrous comedy *All These Women,* all of them are contemporary; most of them take place in "Bergman country," the Swedish countryside or coast seen in haunted and haunting light; all of them have small casts and the effect of spiritual microscopy. With quiet seriousness, Bergman has addressed the largest questions that can still be asked: questions about faith, wholeness, love as an idea and love as practice, hope, the persistence of the beast, the teasings of truth—of the possibility of truth. As Bergman has reduced his means and distilled his thought, he has also moved toward this continuity of collaborators and terrain as part of the process of distillation, so that artistic means *in themselves* would help the sense of concentration, and would carry implicit references. The very face of Max von Sydow reminds us of his pilgrimages, in other Bergman characters, through related crises.

The original title of this new film was *A Passion.* The change for America is a double distortion because Anna is not the protagonist and because it makes the word "passion" sexual, where it was presumably intended to have its religious meaning. The new film's place as part of the Bergman continuum is clear throughout and is emphasized at the end. As the protagonist disappears from view, the narrator's voice says: "This time he was called Andreas Winkelman."

Andreas (played by von Sydow) has separated from his wife and now lives alone with his books and daily chores on that familiar Bergman island. We never learn what his past work was. Nearby is the country place of a successful Stockholm architect (Erland Josephson) and his wife, Eva (Bibi Andersson). Their house guest is Anna (Liv Ullmann), a widow who was driving the car in which her husband and small son were killed. In the course of time, Andreas has a brief but affectionate affair with Eva, who loves her husband but simply lacks the wholeness for fidelity. (It is gently hinted that the husband knows of this affair; he certainly knows about a past one.) Then Anna, the widow, comes to live with Andreas and, through a year or so, they have some tenderness and some fights. Toward the end they have a violent quarrel in which he beats her.

I omit many, many fascinating details, which are not so much details of a story as of four lives. But one element essential to mention here is a killer on the island, a slaughterer of animals. We get a distant glimpse of this man very early, when he snares Andreas' dachshund in a noose. Then eight sheep are killed one night, and the frightened islanders assault a lonely old man because of it, simply because he was once in a mental hospital and they must have a culprit. Despondent, the old man hangs himself. After the suspect's suicide, another farmer's barn is set on fire and his horse killed.

Andreas goes to this fire just after his fierce fight with Anna. She comes to call for him and, as they drive home, they talk calmly about separating. Then he asks why she came to the fire to fetch him, and she replies, "I came to ask your forgiveness." After he has beaten her! She drives off the road (a small parody of the accident she was in earlier), and the car halts. Andreas gets out, and she drives on.

Then comes the very last scene, a long shot of Andreas pacing back and forth; the camera slowly zooms in, the texture gets more and more grainy and white, and as the screen blanches out, we are just able to see Andreas fall to his knees. All during the slow zoom, we hear the ticking of a clock. This is a reference to the clock we heard much earlier when Andreas read a letter he had taken from Anna's pocketbook: a letter from her (dead) husband which tells the unhappy truth about the marriage that she boasts of as happy.

This last scene sums up a good deal of the film's concerns. Andreas is a withdrawn man with a strong streak of violence in him, who thought he had (literally) beaten his way out of an involvement that had come to chafe him, but who finds that he has been accepted, by Anna, in the truth of what he is—which frightens him. Yet has Anna accepted the *truth?* The ticking reminds us of the letter, of Anna's power of self-deception, that she doesn't lie in the ordinary sense but has a terrible capacity to convert unpleasantness to fit her need for affirmation. As the film finishes, Andreas' passion does not finish; he sees that knowledge of this propensity in Anna will not free him of

her, any more than knowledge of his own weaknesses has freed him
of his old self—despite his island retreat in a rite of self-purification.
Purity and consistency are not possible; to live is to contradict yourself.
To be aware of the contradictions and to bear them *without resignation*
is the final passion.

But this film is not a fable leading to a moral. Its real purpose is to
exist: to present some lives for a time, sounding chords by com-
binations of elements within and between each of them—and using
the violence on the island as a sounding board behind them. The fact
that the film has pulse but no plot, motion forward but no neat conclusion, emphasizes that it is part of the Bergman continuum: a self-
contained chapter, still a chapter.

At different points Bergman brings in each of his four main actors
as himself. Each is seen "backstage" and is asked for comments on the
role he (or she) is playing. At first I was dismayed by this device—
now widely used—to include the making of the film as part of the film,
and I'm not yet convinced that it was entirely beneficial. But at least
(unlike similar efforts by, say, Vilgot Sjöman) the actors' comments
are pertinent and, in themselves, become dramatic ingredients; besides, these backstage glimpses help the continuum idea by showing
us even more aspects of the Bergman "company."

Three of these four sterling actors, named above, are well-known
to Bergman viewers. The fourth, Josephson, may be remembered as
the baron in *Hour of the Wolf*. Josephson, who is director of the
Royal Dramatic Theater in Stockholm, also collaborated with Bergman
on the script of *All These Women*. (They used a joint pseudonym.)
His performance of the suave cynical architect—reminiscent of the
architect in *L'Avventura*—looks easy until we get our glimpse of
Josephson himself and see the difference.

There are bothersome elements in the film. After their visit to the
innocent old man suffering from the islanders' violence, Andreas and
Anna go home and see a Vietnam atrocity on television; and while
they are watching, a bird flies into the windowpane and has to be
killed. The sequence is pat. The narrator's voice, which obtruded
only once in *Persona,* obtrudes more often here. And there are some
subliminal shots toward the end—like an insect flittering in mid air
after the beating scene—that add little.

But *The Passion of Anna* is lovely, partly because of its color.
All These Women, Bergman's first color film, was garish; in this one,

his cameraman-collaborator, Sven Nykvist, has opened up the dissolved-pearl light of the previous "island" pictures to reveal the gold-green tones that lay behind it all the while. "Opening up" is the mode of this whole film. If it is not modeled into completely satisfying dramatic shape, neither is it amorphous. It *happens.* It feeds a hunger for experience—as baffling as our own but more beautifully rendered because it is in artists' hands, as ours is not.

Catch-22

(*July 4, 1970*)

It begins brilliantly. No music, no sound; the titles flash on a black screen. Soon a dog barks in the distance, and, behind the titles, light begins to outline a far ridge. Another dog, a gull—and dawn breaks in our eyes across a bay. As the titles finish, invisible B-25s grumble to life, huge plane wheels roll past. Through the departing bomber mission, we move across the field to a ruined building where a man is standing. Closer, we see that it is Yossarian and that behind him are Colonel Cathcart and Lieutenant Colonel Korn. They speak, but we can't hear their words over the roar. Yossarian shakes hands with them and leaves. As he crosses the field, a figure in GI fatigues stabs him.

So at once there are some promises. Control. A really fresh idea of how to open. Balance between gravity (dawn blasphemed by the bombers) and wit (the voices we can't hear). Design: the screenplay adapted with a pattern in mind. In the novel, the stabbing occurs near the end, and presumably it has been transposed here as an arc of a circle to be completed. (Which is what happens. And when we see the scene again, later on, we hear the dialogue that was obscured the first time.)

But most of the promises are broken. *Catch-22,* probably the most eagerly awaited film of a novel since *Gone with the Wind,* is a disappointment. Partly this is because Joseph Heller's novel is now a historical work. Not because it follows *How I Won the War,* the best antiwar film since *Grand Illusion,* or the recent and much lesser

M*A*S*H, but because there has been a profound cultural shift in the nine years since *Catch-22* appeared. Retrospectively we can see that Heller's novel itself made a great cultural sweep: it brushed away *The Catcher in the Rye* as the sensibility banner of the age. The Salinger ideal of private fineness, rather proud to be private and unappreciated by the world, was replaced by the Heller view of society as a prison run by oafs and madmen for whom the few sane inmates felt some weird kind of compassion. Since 1961 no other novel, American or otherwise, has supplanted *Catch-22* as tonal center, which says something about the novel form as such because Heller's book *has* been supplanted—by a combination of rock and psychedelia and films and astrology and social activism. The result is a very different world-outlook. Yossarian had to fly fifty missions before he saw that "the enemy is anybody who's going to get you killed, no matter *which* side he's on. . . ." Today's young men—more liberated, less addicted to disillusion, in some ways precociously mature—know before they get into a uniform what Yossarian had to learn the hard way.

Still, to say that *Catch-22* is now a work in history is certainly not to say that it is dead. For one thing, Yossarian was one of the teachers of present-day youth. For another, in the year when our President says he widened a war in order to hasten peace, the logic of this novel is hardly unrecognizable. The book could have been filmed effectively if it had been treated competently as what it is, without nervousness, if its tone had been caught. For me, that tone is neither savage Swiftian satire nor Manichean black humor but the cool extension of the horror and the ridiculousness that are already present in the world: as in Kafka. (Kafka thought his novels were funny.)

But the film consistently bumbles that tone—or any other tone. In the book, when U.S. planes bomb their own base as a service to the Germans, so that in return the Germans will buy the Egyptian cotton which Minderbinder's syndicate has overstocked, it is a tart, peripheral, insane note; in the film it is an air raid, with real explosions and real flames. When Yossarian walks desolate through the nighttime streets of Rome, Heller makes it a floating expressionist experience. Here it is a solitary parade past a lot of arbitrary symbols—with Donizetti on the soundtrack. In the book, when the parents and brother of a supposedly dying soldier come to visit him and Yossarian substitutes for the boy (who is already dead), it is a scene of shadows and bandages. Here it is played in light, no bandages, and the visitors are three comic actors. A moving symbolic moment becomes a failed skit.

I'm not going to recount the story of the best-read American novel of the last decade. Buck Henry's screenplay omits some characters and collapses some of the others together, predictably. The best element is the retention of Heller's "recurrence" pattern: principally the scene between Yossarian and the wounded Snowden in the shattered plane, which tells us a bit more of the incident each time we return.

And cinematically those Snowden scenes are the best: overexposed to a ghastly whiteness, with chilly wind whistling through them. Some other sequences are well done: Snowden's funeral, with naked Yossarian in a tree; a *trattoria* scene in the Piazza Navona that catches the enforced itchy camaraderie of soldiers on leave; a shot of Doc Daneeka sitting alone on the beach at night. Some of the performances are good: Jon Voight as Minderbinder, Chuck Grodin as Aardvark, a voluptuous unpretty Italian girl named Olympia Carlisli as Luciana, Robert Balaban as Orr, and Art Garfunkel (of Simon &) as Nately. And now I know what I have been waiting for Anthony Perkins to play: not a mad murderer or Melina Mercouri's lover but an Anabaptist chaplain.

But the shocking fact about this film directed by Mike Nichols is that some of the casting is unwise and some of the performances are poor. Balaban and Garfunkel are confusing look-alikes. Martin Balsam (Cathcart) behaves like a prosperous Brooklyn butcher. Bob Newhart (Major Major) is a forlorn disaster. Orson Welles, who never had much comic gift, is highly uncomfortable as Dreedle, and Austin Pendleton caricatures his ninny son-in-law. It was a nice linkage to use Marcel Dalio, who was in *Grand Illusion,* but Dalio overdoes the brothel ancient and, playing an Italian, has an unmistakable French accent. And the picture founders on the one performance that was needed to save it: Alan Arkin's Yossarian.

Arkin has superfine comic timing, which he showed in Nichols' New York production of *Luv.* So he can make us laugh when he discovers in flight that his parachute is missing. He has some sense of pathos. So he is attractively quiet in the bed scene with Luciana. But he is extremely limited technically—Martin Sheen, the Dobbs, might have taught him how to laugh, to name one elementary matter—and he smears his poor speech, rather proudly, over everything. When Arkin hears that Orr has escaped, he exclaims, "Eemayih?" This means "He made it?" And the two harrowing moments, Nately's death and the discovery of Snowden's second wound, are far beyond Arkin's grasp. Yossarian's words in the latter scene—"There, there"—are movingly inadequate in Heller's novel. In Arkin's mouth they seem unlikely and inexpressive. The very same film, shot for shot, would have been immensely improved with a good Yossarian. Heller's hero is the son of Schweik, emigrated to America and immensely wised up. Arkin's Yossarian is not a resolutely sane man who, by his sanity, seems mad; he is a Jewish-intellectual cabaret comic wallowing in antiheroism as he plays for predictable coterie responses.

The cinematographer was the very gifted Englishman David Watkin, whose career includes—hardly by coincidence, I should think—such antiwar films as *The Charge of the Light Brigade, How I Won the War,* and *The Bed Sitting Room.* Some of his work here, like the beginning and the Snowden sequences and a telephoto shot of a bomber-group take-off, is exquisite. (One fantasy scene, of soldiers lined up outside a brothel, is like a waiting line of girls in *The Knack,* also done by Watkin.) But the director hasn't asked him for enough in such scenes as Yossarian's Roman walk, has called for repetitious knee-level shots, and has allowed some fruity lighting, as in the Dalio scene.

And so to the director. I have thought of Mike Nichols as a man who was being punished, by a species of critical Calvinism, because he had been a member of a satirical duo and had directed several frothy comedies on Broadway. The fact that he had directed them with immense skill did not impress his detractors; to them, anyone who accepted those assignments was tainted. (A queer distinction between theater and film; bevies of film critics keep telling us that film material matters little, execution is all.) For myself, I tremendously admired his work in, say, the New York production of *The Knack,* and I thought that his first film, *Who's Afraid of Virginia Woolf?,* was a largely successful struggle. It was a tough first assignment, to have to transfer a distinguished play to the screen, but Nichols showed some cinematic intelligence plus his already celebrated skill with actors. In *The Graduate* he (rightly) opted for more malleable material and more of his self came through; the picture was marred by facetiousness and script weaknesses, but it had the same superb skill with actors, a firmer grip of cinematic influences, and the dynamics of a film whose time has come. *Catch-22* is, on balance, his poorest film to date. Basically, I think this is because he reverted to the strictures of Film One, taking on a distinguished work and again becoming essentially a transposer. (I'm speaking solely from the viewpoint of a potentially superior director; the question of who ought to film works like Albee's play and Heller's novel is another subject.) And this time the original is very much harder to transpose.

Throughout *Catch-22* there are suggestions that Nichols felt trapped; there is a scent of panic. Possibly this accounts for his failure to fix the tone and for his uncharacteristic mistakes with actors. He seems to have realized too late the enormous difficulties of filming this book, of conveying its cosmos within a reasonable length, of making *visible* its understated lunacies, of dealing with its changed position in our culture. So he seems, figuratively, to have ad-libbed his way through the picture, sometimes hitting a good note, sometimes not, and often falling back, when insecure, on revue gimmickry to see him through. (Arkin's Second City repertoire. An "in" reference to *2001*: with the first close-ups of Luciana's body, we hear the Richard Strauss music that began the Kubrick film.) Underneath the film one can almost hear a voice of strained *noblesse oblige* saying: "I am Mike Nichols, famed for social and political wit, and therefore I must do *Catch-22*." Then, sotto voice: "Help!" The paradox is that a lesser director, less oppressed by perception of the difficulties, might have made a more satisfactory picture, without Nichols' high spots but also without his thrashing about.

"Character is destiny," Heraclitus said, freezingly, 2,500 years ago. But what is Nichols' character? After enumerating the good and bad sequences in *Catch-22*, one is left with a balance sheet rather than a man. His career has shown thus far that, at whatever level, he was born to direct. In the theater he has chosen to meet few challenges except those of craft. In films, where optimally the director is at least as much creator as interpreter, he has chosen, two out of three times, the dubious refuge of previously successful works; and this last time, the refuge became a sort of prison, within which he is furtively and almost anonymously clever. What destiny has Nichols' character in store for him? More hot properties? More juggling? He says that he is now just ready to begin, to make his "statement." I hope so. I think that out of the often dazzling gifts a recognizable and valuable man might emerge. I hope that he is not destined for one more of those American careers in art, a man merely living by his talents—which is about the easiest way that a talented man can live.

How Young You Look

(June 27, 1970)

STILL more Now films. They come so thick and fast—no, thin and fast—that there's no point in saying it's "time" for reassessment. Last month was one "time"; next month will be another. But a clutch of three recent films, cited below, are more interesting to discuss generally than to review.

A prime mistake about the Now or Youth film is the assumption that youth-worship is new in this country. It has been around a long while. Slick magazines always had a heavy preference for stories about people under thirty, and so did radio shows. Leo McCarey's film *Make Way for Tomorrow* (1937), a better tearjerker than many successful ones, was a flop because it dealt with old folks. Young people's lack of interest in their elders, as subjects of film or fiction, was only a trifle greater than older people's lack of interest in themselves. But—the crucial difference—older people used to control the youth image, and they shaped it to suit themselves: young folks in those films might stray or rebel, but sooner or later they saw the wisdom of their elders' ways. Until quite recently the guideline for youth in films, invisible but dominant, was: "When I was sixteen, my father was pretty dumb. Now that I'm twenty-one, he's a lot smarter. Amazing how much the old man learned in five years." It made everybody beam.

Today, of course, the ending reads: "Now that I'm twenty-one, he's even dumber. He'll never learn."

In late 1967, with *The Graduate,* the control of the youth image by the non-young was cracked. The crack came about, principally, because of an altered moral climate and the downward age shift in the population. There were proportionately more young people than ever, and they were more grievously troubled than ever, and the first picture to acknowledge that truth with any degree of seriousness was a torrential success. The subsequent pictures that flooded through the crack, as I have pointed out many times, have varied widely in quality, and none of them has been first-rate. American film art has not zoomed to Olympus. But a large basic change has taken place: American films no longer need to use codes or disguises or patronizing transformations in order to treat contemporary subjects. At least there is now much more verisimilitude, even if—so far—there is not much serious art.

Now, after the recent *Getting Straight,* there are three more Youth films. *The Magic Garden of Stanley Sweetheart* (MGM) is about the sexual and drug adventures of a Columbia undergraduate. *The Landlord* (United Artists) is about a rich twenty-nine-year-old who buys a run-down house to remodel it for himself, discovers that it is in a black ghetto and has black tenants, and learns about life. (A faint echo here of Shaw's *Widowers' Houses.) The Grasshopper* (National General) is about a twenty-year-old girl from British Columbia, as eager for adventure as ever Joan Crawford was but much less inhibited, who becomes a ripple on life's stream in Las Vegas and Los Angeles. These three films share three sets of characteristics. First, they all have broad language, nudity, frank sexual references, casual use of drugs, and miscegenation without comment. (That is, the mixed couples never mention color as a possible barrier to bed.) Second, they all have several songs on the sound track reminiscent of Simon and Garfunkel. Third, they all use plentiful up-to-date camera and editing techniques.

Verisimilitude is the key word. These films look and sound more like American life than American films used to. But their handling of materials that were recently taboo is so glib that we are never even faintly deluded about their sincerity. The sound-track songs are so obtrusive that these films almost seem like devices to promote record albums. And the cinematic display is piled on so mountainously that it dwarfs the already tiny content. For instance, there is a bed scene between the hero and his girl in *The Landlord* that is filled with gargantuan soft close-ups and lavish lyric lighting, utterly disproportionate to the two characters or even to the event's importance to *them.* The new directors (all three of these men are fairly new) have film tastes shaped less by good film models than by the TV commercial, which packs as much cinematic display as possible into those sixty seconds.

So, in less than three years, the Youth or Now or Mod film has solidified into a staple. The ostensibly "personal" film has become one more line of goods, once again "integrity" has become a prop of salesmanship, once again the Bitch Goddess feeds happily on her ostensible critics. Hollywood—still a usable term even if now less accurate geographically, even if now staffed with lots of university graduates—used to be cynical about lies; now it is cynical about truth.

It might seem, then, that those of us who took some pleasure in the American film's widened compass have already been bilked. But, if I'm not exactly happy about developments, neither am I depressed. Or surprised. As between cynicisms, I prefer the new to the old, and since it is the nature of Hollywood to be cynical, the choice is precisely between cynicisms. When serious subjects became admissible seriously, did we expect them to be banned to all but serious film makers?

More important, the territory of the American film *has* been widened. That truth may prove irreversible. Some verisimilitude, at least, of the American psyche is now a commonplace on the American screen. In a popular-based art like the film, the figurative shape of the art is pyramidal: the broader the base, the higher the possibilities. If there had been no superficially probing Clyde Fitch or Augustus Thomas, Eugene O'Neill's job would probably have been much harder. An unconstricted Swedish film culture produces, along with its many sexual and psychologizing bores, an Ingmar Bergman. The new American film territory at least makes serious American film art more possible.

The former Youth image, controlled by the non-young, flattered the non-young. The new Youth image, controlled by the young, flatters the young. Like so many changes brought about by the young these days, it is at least partially an improvement. And like so many such changes, it is up to the young not to be pleased with partial success or to overlook what has been left undone or badly done, or to be co-opted by patronization. Hollywood, still cynical, will dance to any tune if enough people call it. The young audience—which is not only the majority audience now but on the whole the best film audience this country has ever had—can call better ones. If only it will get over its childish (not young!) pleasure in having been recognized, if only it will demand the right to be angered, to be expanded. Most Youth films so far are "recognition" films: they get their effect by providing accurate correspondence with the world outside. But that is where the best films *begin*.

Five years ago, in an essay called "The Film Generation," I called that generation "the most cheering circumstance in contemporary American art." This seems to me to be even more clearly the case now. I also hoped that the Film Generation would make some demands. It has, and now it is in control. Therefore it is now in danger. The audience rules the film world, and rulers are always the target of flatterers.

the necessary film

When the first moving picture flashed onto a screen, the double life of all human beings became intensified. That double life consists, on the one hand, of actions and words and surfaces, and, on the other, of secrets and self-knowledges, self-ignorances, self-ignorings. That double life has been part of man's existence ever since art and religion were invented to make sure that he became aware of it. In the past century, religion has receded further and further as revealer of that double life, and art has taken over more and more of the function; and when the film art came along, it made that revelation of doubleness inescapable, more *attractive*. On the screen are facts; which at the same time are symbols; and they thus invoke doubleness at every moment, in every kind of picture. They stir up the concealments in our lives, both those concealments we like and those we

281

don't like, they shake our histories and our hopes into our conscious-ness. Not completely, by any means. (Who could stand it?) Not more grandly or deeply than do other arts. But more quickly and surely.

Think of this process as applying to every frame of film and it is clear that when we sit before a screen, we run risks unprecedented in human history. A poem may or may not touch us, a play or novel may never get near us. But films are inescapable. (In fact, with poor films, we often have the sensation of fighting our way *out* of them.) When two screen lovers kiss, in any film, that kiss has a minimum inescapability which is stronger than in other arts—both as an action before us and as a metaphor of the kissingness in our lives. Each of us is pinned privately to that kiss in some degree of pleasure or pain or enlightenment. In period films or modern dramas, in musicals or political epics, in Westerns or farces, our beings are in some measure summoned up before our private vision.

And I suggest that the fundamental way, conscious or not, in which we determine the quality of a film is by the degree to which the re-experiencing of ourselves coincides with our pride, our shames, our hopes, our honor. Finally, distinctions among films arise from the way they please or displease us with ourselves: not *whether* they please or displease but *how*.

This is true, I believe, in every art today; it is not a cinema mo-nopoly. But in film it is becoming more true more swiftly and deci-sively because the film has a much smaller heritage of received esthetics to reassess; because the film is bound more closely to the future than other arts seem to be; and because the film confronts us so im-mediately, so seductively, and so shockingly with at least some truth about what we have been doing with ourselves.

To the degree that film exposes a viewer to this truth of himself, in his experience of the world or of fantasy, in his options of actions or of privacy, to the degree that he can thus accept a film as worthy of himself or better than himself, to that degree a film is necessary to him; and it is that necessity, I suggest, that ultimately sets its value.

Throughout history, two factors have formed men's taste in any art—knowledge of that art and knowledge of life—and obviously this is still true, but the function of taste seems to be altering. As formalist esthetic canons seem less and less tenable, standards in art and life become more and more congruent, and the function of taste seems increasingly to be the selection and appraisal of the works that are most valuable—most necessary—to the individual's *existence*. So our means for evaluating films become more and more involved with our means for evaluating experience: not identical with our standards in life but certainly related—and, one hopes, somewhat braver.

Of course the whole process means that men feed on themselves, on their own lives variously rearranged by art, as a source of values. But despite other prevalent beliefs in the past, we are coming to see that men have always been the source of their own values. In the century in which this liberation, this responsibility, has become increasingly apparent, the intellect of man has simultaneously provided a new art form, the film, that can make the most of it.

index

Page numbers referring to principal discussion of a film are set in *italic*.

285

ABOUT THE AUTHOR

Stanley Kauffmann has been writing film criticism for *The New Republic* since 1958 and is at present the film and drama critic. He has also served as drama critic for *The New York Times*.

Mr. Kauffmann, a native of New York, is a graduate of New York University.

71 72 73 10 9 8 7 6 5 4 3 2 1